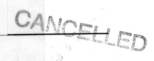
A Beginner's Guide to Mass Spectral Interpretation

A Beginner's Guide to Mass Spectral Interpretation

Terrence A. Lee
Middle Tennessee State University
Department of Chemistry
Murfreesboro
TN 37132
USA

JOHN WILEY & SONS

Chichester • New York • Weinheim • Brisbane • Singapore • Toronto

Copyright © 1998 John Wiley & Sons Ltd,
Baffins Lane, Chichester,
West Sussex PO19 1UD, England

National   01243 779777
International (+44) 1243 779777
e-mail (for orders and customer service enquiries): cs-books@wiley.co.uk
Visit our Home Page on http://www.wiley.co.uk
          or http://www.wiley.com

Other Wiley Editorial Offices

John Wiley & Sons, Inc., 605 Third Avenue,
New York, NY 10158-0012, USA

WILEY-VCH Verlag GmbH, Pappelallee 3,
D-69469 Weinheim, Germany

Jacaranda Wiley Ltd, 33 Park Road, Milton,
Queensland 4064, Australia

John Wiley & Sons (Asia) Pte Ltd, Clementi Loop #02-01,
Jin Xing Distripark, Singapore 129809

John Wiley & Sons (Canada) Ltd, 22 Worcester Road,
Rexdale, Ontario M9W 1L1, Canada

Library of Congress Cataloging-in-Publication Data

Lee, Terrence A.
    A beginner's guide to mass spectral interpretation   /   Terrence A.
Lee.
        p.   cm.
    Includes bibliographical references (p.   –   ) and index.
    ISBN 0-471-97628-8 (hardcover : alk. paper).—ISBN 0-471-97629-6
(pbk. : alk. paper)
    1. Mass spectrometry. I. Title.
QD96.M3L44 1998
547'.3'0873—dc21
                                                                97-28548
                                                                CIP

British Library Cataloguing in Publication Data

A catalogue record for this book is available from the British Library

ISBN 0 471 92628 8 (cloth)
ISBN 0 471 97629 6 (paper)

Typeset in 10/12 pt Times Roman by Techset Composition Ltd, Salisbury, Wiltshire.

Printed and bound in Great Britain by Bookcraft (Bath) Ltd

This book is printed on acid-free paper responsibly manufactured from sustainable forestry, in which at least two
trees are planted for each one used for paper production

Dedication

This book is dedicated to Joyce Jamil, Myron Jones and Nathan Hurt, who as students at Middle Tennessee State University provided the inspiration to write this manual. In addition to the students who inspired this book, the author also dedicates this book to Lisa A. Lee and Krista D. Lee, my wife and daughter, who have encouraged and supported all of my efforts.

Contents

α, alpha cleavage	The breaking of a bond to an atom adjacent to the atom containing the odd electron (not the bond to the atom containing the odd electron).
'A' element	An element that is monoisotopic.
'A + 1' element	An element with an isotope that is 1 amu above that of the most abundant isotope, but which is not an 'A + 2' element.
'A + 2' element	An element with an isotope that is 2 amu above that of the most abundant isotope.
amu	Atomic mass units. For our purpose, the sum of the protons and neutrons in the nucleus ($^{12}C = 12$ amu).
base peak	The peak (or line) in the spectrum that represents the most abundant ion.
daughter ion	The product produced by some sort of fragmentation of a larger ion (see parent ion, below).
EE^+, even-electron ion	An ion in which the outer electrons are fully paired.
EI	Electron ionization (or impact).
eV	Electron volts
isotopic peak	A peak (or line) in the spectrum that corresponds to the presence of one or more heavier isotopes in an ion.
m/z	The mass of an ion divided by the electrical charge of the ion. Normally, the charge is $+1$. Sometimes m/e is used synonymously.
molecular ion, $M^{+\cdot}$	The ionized form of the molecule. The 'molecular ion' is the peak (or line) in the spectrum corresponding to a molecule that is composed solely of the most abundant isotope forms. Sometimes 'parent ion' and 'parent peak' are used. Frequently, 'M', 'M + 1', 'M + 2', etc. will be used to indicate the molecular ion and isotopic peaks.
n-electrons	Non-bonding electrons
$OE^{+\cdot}$, odd-electron ion	An ion with one unpaired electron in the outer shell. Also known as a radical ion.
parent ion	The ion which is decomposing or breaking apart.
π-electrons	Electrons in double or triple bonds, or in aromatic ring systems such as benzene.
R + DB	The number of rings and double bonds. This is a measurement of the saturation of a compound. Completely saturated compounds (such as propane) will have 0 for the R + DB.
relative abundance	The abundance of a given ion relative to the base peak. Usually, the base peak is set to 100%.
σ, sigma cleavage	A simple cleavage reaction taking place by the ionization and breaking of a sigma bond.
simple cleavage	A reaction which involves the breaking of one bond.
	Movement of a pair of electrons.
	Movement of a single electron.

1 Isotopic Abundances and How We Use Them

The chemical elements making up our world exist in several different forms, known as isotopes. The term 'isotope' is used to describe atoms of an element with varying numbers of neutrons. Carbon occurs in two naturally occurring isotopic forms: ^{12}C, having six protons and six neutrons, and ^{13}C, having six protons and seven neutrons. Most of the elements in the Periodic Table will have two or more isotopic forms. Some of these isotopes are radioactive and some are non-radioactive. The non-radioactive isotopes are known collectively as 'stable isotopes'. All isotopes of an element are chemically identical. They participate in identical chemical reactions and share identical chemical properties, with the exception that some reaction rates (or properties) will vary owing to differences in mass.

One of the simplest examples of how mass can affect the rate of a process is to look at the effusion of uranium hexafluoride (UF_6) through a porous plate. Uranium exists predominantly as two isotopes, ^{235}U and ^{238}U. Graham's law of effusion tells us that the relative rates of effusion for two gases at the same temperature are given by the inverse ratio of the square roots of the masses of the gas particles. Mathematically

$$\frac{\text{rate 1}}{\text{rate 2}} = \frac{(\text{mass of gas 2})^{1/2}}{(\text{mass of gas 1})^{1/2}}$$

where the superscript 1/2 represents taking the square root of the mass of the gas.

The molecular weight of $^{235}UF_6$ is approximately 349 amu, and that of $^{238}UF_6$ is approximately 352 amu. If we assign (arbitrarily) $^{235}UF_6$ as gas 1, then the ratio of the square roots of the masses is

$$18.76/18.68 = 1.004$$

which indicates that the isotope with the lower mass will effuse at a faster rate than the higher mass isotope. This higher rate amounts to about 0.4%, which is a typical isotope effect.

What about the magnitude of this isotope effect for smaller organic molecules? Well, a similar calculation can be made for a molecule such as benzene. The molecular weight of a benzene molecule made from only ^{12}C and ^{1}H would be 78 amu. The molecular weight of a benzene molecule containing one ^{13}C atom would be 79 amu and the ratio of the square roots of these masses would be

$$8.88/8.83 = 1.006$$

and there would be approximately a 0.6% difference in effusion rate.

The largest isotope effect is normally seen when significant amounts of ^{2}H (deuterium) are present in a relatively small molecule. If we make a similar calculation to those above comparing ordinary water with monodeuterated water, we would see that the ratio is 1.027 or there is an approximate 2.7% difference in the effusion rate. This is a fairly extreme example of an isotope effect. For most applications, the isotope effect is insignificant and can be ignored.

Naturally occurring isotopes, either stable or radioactive, do not occur in equal amounts. Naturally occurring ^{13}C has an abundance of about 1.1%; i.e. out of 1000 carbon atoms chosen at random, about 11 of them will be the heavier ^{13}C isotope and the other 989 will be ^{12}C. Although ^{14}C occurs in nature, the natural ^{14}C abundance is so low as to be undetectable by ordinary mass spectrometric methods. It is fortunate that, for the elements commonly found in organic compounds, the heavier isotopes are lower in abundance. The natural isotopic abundances of common elements found in organic compounds are given in Table 1.

By convention, we normally designate oxygen, sulfur, silicon, chlorine and bromine as 'A + 2' elements. This is because the important isotopes of these elements differ from

Table 1 Mass and relative abundance of common organic elements

Elements containing only one important isotopic form	
Element	Mass
H(A)	1
F(A)	19
P(A)	31
I(A)	127

Elements containing two important isotopic forms				
Element	Mass	% Abundance	Mass	% Abundance
C(A + 1)	12	100	13	1.1
N(A + 1)	14	100	15	0.37
Cl(A + 2)	35	100	37	32.5
Br(A + 2)	79	100	81	98.0
O(A + 2)	16	100	18	0.20[a]

Elements containing three important isotopic forms						
Element	Mass	% Abundance	Mass	% Abundance	Mass	% Abundance
Si(A + 2)	28	100	29	5.1	30	3.4
S(A + 2)	32	100	33	0.80	34	4.4

[a] Oxygen-17 is found at about 0.04% natural abundance. The percentage abundances listed for the other isotopes are normalized to the most common form.

each other by two mass units. Carbon and nitrogen are designated as 'A + 1' elements because their isotopes differ by one mass unit. Elements having no important naturally occurring isotopes are designated as 'A' elements, and include hydrogen, fluorine, phosphorus and iodine.

The appearance of isotopically shifted lines in a mass spectrum provides the analyst with information about the molecular formula of the compound or ion that corresponds with the lines. One of the earliest steps in interpreting a mass spectrum is to determine if any A + 2 elements are present. In the case of Br or Cl this can normally be accomplished by inspecting the high mass region of the spectrum, looking for the characteristic isotope patterns produced by these elements. Compounds containing a single chlorine atom will exhibit a pair of lines, separated by 2 amu and in a relative abundance ratio of about 3 : 1. Compounds containing a single bromine atom will also exhibit a pair of lines separated by 2 amu but the ratio will be almost 1 : 1. Similarly, Si or S can be detected by means of their isotopic pattern.

When more than one atom of these A + 2 elements is present in a molecule, the spectrum will become more complex. Clusters of lines will be present at intervals of 2 amu and the spectra of compounds containing multiple Cl or Br atoms (or Cl and Br together) are more complex. The following isotope patterns (Figure 1.1) produced by Cl, Br, Si and S are the most important because they are those most often encountered.

Methods for determining the number of halogens and their identities in polyhalogenated compounds will be presented in Chapter 5.

For the purpose of identifying an organic compound, the most important of the A + 1 isotopes is ^{13}C. The presence of this isotope at an abundance of about 1.1% allows the analyst to calculate the number of carbons present in a molecule or ion fragment. Suppose we have two lines, separated by 1 amu. If carbon is present in the molecule (or ion fragment) associated with these lines, then it would be reasonable to assume that the line of higher mass (designated the m + 1 line) would contain ^{13}C, whereas the line of lower mass (designated the m line) would contain ^{12}C. We can divide the intensity of the 'm + 1' line by the intensity of the 'm' line, obtaining the ratio of (m + 1)/m. Dividing this ratio by 0.011 will give us an indication of the number of carbon atoms present in the molecule.

For example, look at the spectrum of methane (Example 1 in Chapter 4). At the high mass end of the spectrum, we observe lines at m/z 17 and 16. The ratio of the intensities of these two lines is 1.6/100 or 0.016. Dividing this ratio by 0.011 gives us 1.4 as a result. This value indicates that we have at least one carbon atom and certainly fewer than two carbon atoms. For another example of this procedure, consider the high mass region in the spectrum of butane (Example 3 in Chapter 4). Here we have lines at m/z 58 and 59 with a ratio of 0.042. Dividing this ratio by 0.011 gives us 3.8 as a result, which indicates the presence of more than three carbons but fewer than five carbons.

Notice that in our two examples the final results were not whole numbers. This is often the case, and a certain amount of care must be used when calculating the number of carbon atoms present in a molecule or ion fragment. A variety of factors can affect the relative intensities of lines in a mass spectrum, and not all lines are due solely to the presence of isotopes.

Although the presence of nitrogen can be determined by its ^{15}N isotope, it is normal to use the nitrogen rule (Chapter 2) to determine whether or not this element is present in a given compound. This is because of the relatively low abundance of ^{15}N (0.37%). Similar

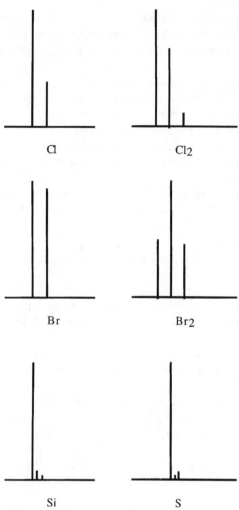

Figure 1

difficulties are encountered with oxygen and the ^{18}O isotope. This $A + 2$ element has an isotopic abundance of 0.20%, and its presence or absence is determined by other methods that will be described in later chapters.

As a general rule, once a compound exceeds about six carbons, the direct determination of nitrogen and oxygen by their respective isotope peaks becomes very difficult. This difficulty is due to the normal isotopic abundance of ^{13}C. For a C_6 compound, the abundance of the $A + 1$ line due to ^{13}C is about 6.6% and that of the $A + 2$ line is about 0.18%. The abundance of the $A + 2$ line is caused by the possibility of having two ^{13}C atoms in the molecule. As the number of carbon atoms in a molecule increases, the corresponding intensities the $A + 1$ and $A + 2$ lines increase, making the direct determination of nitrogen and oxygen from their isotopic abundances extremely difficult or impossible.

2 Identification of the Molecular Ion

The first step (and arguably the most important) in the interpretation of a mass spectrum is the identification of the molecular ion (also known as the parent ion). The molecular ion is the ionized form of the neutral compound. For example, if the molecule we are interested in is methane (CH_4), the molecular ion will be CH_4^+ (a molecule of methane minus 1 electron).

Unfortunately, not all compounds will produce a molecular ion that is stable enough to be seen in a mass spectrum. Many alcohols, esters and carboxylic acids, among others, will not exhibit any significant intensity at the m/z value corresponding to the molecular ion. How then can we determine if a particular m/z value represents the molecular ion?

In order for a particular m/z value to be assigned as the molecular ion, three criteria must be met. If the m/z value meets all three criteria, it may be the molecular ion. If it does not meet all three criteria, it is not the molecular ion.

1. The molecular ion must be the highest mass ion in the spectra, discounting isotope satellites. This is because the molecular ion is the ionized form of the compound. Fragments from this parent compound are responsible for the other lines found in the spectrum of the compound. In the spectrum of a pure compound, it makes little sense to investigate a line at m/z 80 as a potential molecular ion when there is another line at m/z 96.

Allowances are made for isotope effects. For example, in the spectrum of methane, the highest m/z value will be at m/z 17. This is not the molecular ion but instead is due to the ^{13}C form of methane.

Typically, the molecular ion will have an even mass value. Compounds containing only C, H, O, F, Cl, Br, I, Si, P and S will have even molecular weights. Consider the following representative compounds:

Compound	Formula	Nominal mass
Ethane	C_2H_6	30
Octane	C_8H_{18}	114
Ethanol	C_2H_6O	46
Propanoic acid	$C_3H_6O_2$	74
Dichloromethane	CH_2Cl_2	84, 86, 88 (isotopes)
Hexafluoroethane	C_2F_6	138
Carbon disulfide	CS_2	76
Iodomethane	CH_3I	142

All of these compounds will have even masses because of the nature of the elements from which they are composed. Elements having an even atomic mass also have an even valence, and those elements with an odd atomic mass have an odd valence:

Element	Mass	Valence
H	1	1
C	12	4
O	16	2
F	19	1
Si	28 (30)	4
P	31	3
S	32 (34)	2
Cl	35 (37)	1
Br	79 (81)	1
I	127	1

The exception comes with N-containing compounds. This is because, although N has an even atomic mass (14), it has an odd valence (3). This leads to what is described as the nitrogen rule:

> A molecule containing an odd number of nitrogens will have an odd molecular weight, while a compound containing no nitrogens or an even number of nitrogens will have an even molecular weight.

2. The ion must be an odd-electron (OE) ion. Ionization involves the collision of an electron with a molecule, resulting in the production of the molecular ion by removal of an electron:

$$ABCD + e^- \rightarrow ABCD^{+\cdot} + 2e^-$$

A relatively simple example can be seen in the ionization of methane (CH_4). In the methane molecule there are six electrons from the carbon and four electrons from the hydrogens for a total of ten electrons. During ionization one of these electrons is removed, and the resulting molecular ion has nine electrons.

The most convenient way of determining whether or not an ion is an odd-electron ion is by calculating the saturation index. The saturation index is simply the number of rings and/or double bonds (R + DB) that a molecule possesses and is calculated from the molecular formula. The saturation index is found by the following calculation.

For the general formula $C_xH_yN_zO_n$:

The total number of rings + double bonds $= x - 1/2y + 1/2z + 1$

Note that the number of oxygens present plays no part in calculating the saturation index.

If there are elements present in the molecular formula that are different from the four listed above in the general formula, then:

Si is treated as if it were C
P is treated as if it were N
S is treated as if it were O
F, Cl, Br and I are treated as if they were H

Notice that the valence of the elements plays an important role in this calculation. H, F, Cl, Br and I all have a valence of 1, Si and C have a valence of 4, P and N have a valence of 3 and S and O have a valence of 2.

Let us look at some examples to make sure that the process of calculating the saturation index is clear. Consider the methane molecule, CH_4. Calculation of the saturation index gives

$$1 - 1/2(4) + 1 = 0$$

Since 0 is a whole number, this indicates that this ion is an odd-electron ion and therefore could be the molecular ion. All odd-electron ions will result in saturation indices that are whole numbers. The result also indicates that there are no rings or double bonds present in the molecule.

Consider benzene (C_6H_6). Calculating the saturation index gives us

$$6 - 1/2(6) + 1 = 4$$

once again resulting in a whole number. Benzene contains three double bonds, forming one ring.

What happens when oxygen is present? Let us perform the calculation on benzoic acid ($C_7H_6O_2$).

$$7 - 1/2(6) + 1 = 5$$

Once again we would have an odd-electron ion containing a total of five rings plus double bonds. A glance at the structure of benzoic acid (Figure 2.1) would show that it does indeed contain a total of five rings plus double bonds; three C=C, one C=O and the benzene ring.

Another example. We determine that a reasonable formula for a particular m/z value would be C_7H_5O. The saturation index is:

$$7 - 1/2(5) + 1 = 5.5$$

Since the saturation index is not a whole number, this indicates that the ion is an even-electron (EE) ion and cannot be the parent. This is an important characteristic of even-electron ions—they will never have whole number values for their saturation index. It also indicates that in this ion there is a total of five rings plus double bonds. It does not tell us how many of each (rings and double bonds) are present, although we may be able to propose a structure for this ion if we know the molecular formula of the parent compound.

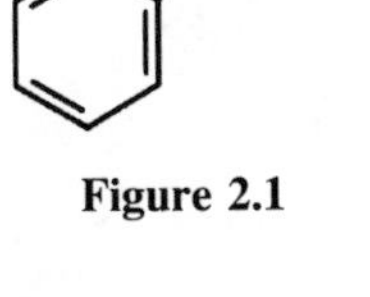

Figure 2.1

Consider the formula for pyridine, C_5H_5N. Calculation of the saturation index gives us

$$5 - 1/2(5) + 1/2(1) + 1 = 4$$

Since this value is a whole number, this ion is an OE ion and could be the molecular ion. This ion also has a total of four rings plus double bonds. Note in this case that the mass of pyridine is 79. This is an odd mass but is allowable because of the nitrogen rule.

Students are advised to use this procedure with caution. While it is true that all molecular ions will be odd-electron ions, not all odd-electron ions are molecular ions. Many compounds can form odd-electron ions by breaking two chemical bonds. The most common type of reaction producing an odd-electron ion that is not the parent ion is the McLafferty rearrangement. In this reaction, a small neutral molecule is ejected from the parent compound. The neutral molecule will commonly have an even mass. Specific examples of this type of reaction will be discussed in later chapters.

One additional word of caution. From time to time, the student will calculate a saturation index with a negative value. This negative value indicates that the molecular formula is incorrect, and the most common cause is that there are not enough carbons present in the proposed molecular formula. If this happens, the student should increase the number of carbon atoms by 1, generate a new molecular formula and re-calculate the saturation index.

3. The compound represented by the molecular ion must be capable of producing the important ions in the 'high mass' region of the spectrum. High mass is of course a relative term, and the spectrum of a compound with a molecular ion at m/z 60 will have a different high mass region from a compound with a molecular ion at m/z 200. We will expend a good deal of effort on this topic in later chapters.

While the calculation of the molecular formula is a very important step in the interpretation of a mass spectrum, an equally important step is the determination of the structural formula. The exact arrangement of the atoms will have significant effects on the appearance of the mass spectrum. A correct molecular formula reduces the number of possibilities by eliminating immense numbers of chemical compounds that cannot produce a given spectrum, but it cannot indicate the specific structural isomer (except for very simple compounds).

3 General Interpretation Procedures

We are now ready to begin our systematic study of the interpretation of mass spectra. Before we get too deeply involved with some examples, I want to offer some advice and a warning to the student. Probably the single most important fact for the student to keep in mind is that not all compounds can be uniquely identified based solely upon the mass spectrum. Quite often, when an analyst is asked to determine the identity of an unknown compound, several complementary techniques will be used. These techniques include infrared spectroscopy (to identify principal functional groups), proton and carbon NMR (to identify chemically identical hydrogens and to determine connectivity in the carbon skeleton), mass spectrometry (for molecular weight, elemental composition and molecular formula), X-ray crystallography, ultraviolet spectroscopy and a variety of classical wet-chemical techniques including melting point and boiling point determinations, refractive index, density etc.

The best way to approach any type of spectral interpretation problem is with a completely open mind. The analyst must be willing to be led by the data provided without trying to force the spectrum to match preconceived notions of the compound's identity. Successful interpretation involves considering all of the data available, including the identity of reactants and known chemical conditions and reactions.

It is always best to approach the interpretation of spectra using a logical and systematic method. The procedure we will follow includes the following steps:

1. Perform a general inspection of the spectrum. Does it have many lines (indicating the presence of many easily broken bonds) or does it have relatively few lines (indicating very stable ions)?

2. Inspect the spectrum for the presence or absence of any A + 2 elements (Cl, Br, Si or S). From this inspection determine the number and kind(s) of A + 2 elements present.

3. Identify the highest m/z fragment in the spectrum discounting the presence of isotope satellites. Is the m/z of the suspected molecular ion odd or even? If it is odd, and if it is the molecular ion, then the presence of an odd number of nitrogen atoms is indicated. Remember that the vast majority of organic compounds will have even values of m/z for the molecular ion. Insert the appropriate error limits (discussed below). From the A + 1 line, calculate the number of carbons present.

4. From the information provided in steps 1–3, write all possible molecular formulae.

5. Determine whether or not the fragment with the highest m/z is an odd-electron fragment. Only odd-electron ions can be considered as potential molecular ions.

6. Identify the presence of any other significant odd-electron ions. If the molecular ion has an even mass, significant odd-electron ions will also have an even mass. If the molecular ion has an odd mass, significant odd-electron ions will also have an odd mass.

7. Using the isotopic ratio, calculate the formulae for the significant odd-electron fragments and determine their saturation index. You should also calculate the formulae and the saturation index for significant even-electron fragments, although in many cases this is not absolutely necessary.

8. Using all of the available information from the spectrum, plus any information from other spectral methods or from chemical and physical tests of the unknown compound, determine a structure for the unknown compound consistent with the data available.

The mass spectra of organic compounds can be relatively simple, containing only a few lines, or amazingly complicated, with dozens of lines. Fortunately, not every line in a spectrum has to be traced to a particular fragment ion. This is neither desirable nor in most cases possible. As you gain experience with interpreting mass spectra you will learn which lines are important and which lines are less important. As a general guide, you will want to identify the molecular ion (if present), the base peak (which is the most intense line in the spectrum), any odd-electron ions that are not the molecular ion and any other relatively intense peaks present.

We will now look at some examples in order to demonstrate and expand upon the steps given above, specifically steps 2–5. In these examples we will concentrate predominantly on the identification of the molecular ion peak (or at least a peak which could be the molecular ion).

Unknown No. 1

Unknown No. 1 exhibits the following spectral abundances:

m/z	Relative abundance
64	100.0
65	0.9
66	5.0

In order to reflect normal instrumental variations, we need to include error limits in our measurements of relative abundance. Normal error limits are either $\pm 10\%$ relative or ± 0.20 absolute (whichever is the larger).

Inserting these error limits, we see:

m/z	Relative abundance
64	100.0
65	0.9 ± 0.20
66	5.0 ± 0.5

This $A + 2$ pattern indicates that one sulfur atom could be present in the molecule. The pattern is inconsistent with the presence of Si, Cl or Br.

m/z	Relative abundance	S_1
64	100.0	100.0
65	0.9 ± 0.20	0.8
66	5.0 ± 0.5	4.4

The *m/z* 66 abundance would allow up to five oxygens in the compound ($5.5 - 4.4 = 1.1$; $1.1/0.2 = 5$), but this is clearly impossible owing to mass limitations. A logical formula for the compound is SO_2:

m/z	Relative abundance	S	O_2
64	100.0	100.0	100.0
65	0.9 ± 0.20	0.8	0.08
66	5.0 ± 0.5	4.4	0.4

Notice how closely the predicted pattern of isotopic abundances for SO_2 matches the actual abundances. We would expect the relative abundance for *m/z* 64, 65 and 66 to be 100.0, 0.88 and 4.8% respectively. These expected abundances are within the observed abundances taking into account the error limits.

Unknown No. 2

m/z	Relative abundance
58	12.0
59	0.5
60	0.0

Inserting the appropriate error limits and normalizing *m/z* 58 to 100% gives us the following normalized data:

m/z	Relative abundance	Normalized
58	12.0	100.0
59	0.5 ± 0.20	4.2 ± 1.7
60	0.0 ± 0.20	0.0 ± 0.20

From the $A + 2$ (*m/z* 60) abundance it is possible that we have one oxygen in our compound. It is also possible that we do not have any oxygen present (and certainly no other $A + 2$ elements).

From the $A+1$ (m/z 59) abundance and the error limits we have a range from 2.5 to 5.9. In order to calculate the number of carbons present, we divide each of these values by 1.1 and round appropriately. This allows us to have three, four or five carbons. The restriction due to mass (a total of 58) eliminates the five carbon possibility. This gives us the following potential starting points for our molecular formula:

	C	O	H
(1)	3	0	
(2)	3	1	
(3)	4	0	

For (1) the balance of the mass ($58 - 36 = 22$) would be due to A elements, since we have already accounted for the only possible $A+2$ elements. The only important A elements left are F and H (iodine and phosphorus are impossible owing to mass considerations). Why do we not consider nitrogen? From the nitrogen rule, we must have an even number of nitrogens in order to have an even mass. Since the smallest even number of nitrogens is two, and two nitrogens represent 28 mass units, we can conclude that there is no nitrogen present in this material.

For formula 1, we are left with two possibilities: C_3H_{22} and C_3H_3F. Clearly, C_3H_{22} is impossible, since in an ordinary saturated hydrocarbon the number of hydrogens is $2n+2$, where n is the number of carbons. For a three carbon hydrocarbon we would have a maximum of eight hydrogens and thus C_3H_{22} is ridiculous.

The other alternative, C_3H_3F, is possible so we will retain it for the time being.

For (2) the balance of the mass ($58 - 52 = 6$) can only be composed of hydrogens and we arrive at the formula C_3H_6O, which is perfectly reasonable.

For (3) the balance of the mass ($58 - 48 = 10$) leaves us only with hydrogens and we arrive at the formula C_4H_{10}, once again perfectly reasonable.

The next step is to calculate the saturation index to determine which (if any) of the proposed formulae results in a possible molecular ion:

Formula	R + DB	Possible molecular ion?
C_3H_3F	$3 - 1/2(4) + 1 = 2$	yes
C_3H_6O	$3 - 1/2(6) + 1 = 1$	yes
C_4H_{10}	$4 - 1/2(10) + 1 = 0$	yes

Notice that all of the values calculated for $R+DB$ are whole numbers. If any of the results had ended in 0.5 then we would be dealing with an even-electron (EE) fragment. In this case we would have eliminated that possibility from our list of molecular ions, since a molecular ion must be an odd-electron (OE) ion. Remember: all molecular ions will be OE ions, but not all OE ions will be molecular ions.

Can we eliminate any of the possible formulae from further consideration? In order to answer this question, we will have to draw some possible structures. If we cannot draw a possible structure based on the formula and saturation index, then we can eliminate that possibility. In the present example, we find that we can draw reasonable structures for each formula.

$$H-C\equiv C-CH_2F$$

$$H_3C-\overset{\overset{\displaystyle O}{\|}}{C}-CH_3$$

$$CH_3CH_2CH_2CH_3 \text{ or } CH_3\underset{\underset{\displaystyle CH_3}{|}}{C}HCH_3$$

We have extracted the maximum amount of information from this particular group of lines and we still have not identified the unknown. This is not too surprising since all we have really accomplished so far is to determine some reasonable formulae for our unknown. We now have to look at the rest of the fragment lines in the mass spectrum and determine which logical losses from the parent compound would be responsible for these lines.

Let's suppose that in addition to the lines at m/z 58–60 there were lines at m/z 43, 44 and 45:

m/z	Relative abundance
43	100.0
44	3.3
45	0.0

Inserting our error limits we would have:

m/z	Relative abundance
43	100.0
44	3.3 ± 0.3
45	0.0 ± 0.2

From the $A + 2$ line, we have to include the possibility of having one oxygen. From the $A + 1$ line, we definitely have three carbons. But if we have three carbons (mass of 36) then we do not have any oxygen.

With three carbons, we have to account for seven additional mass units ($43 - 36 = 7$). Clearly, only seven hydrogens are possible, resulting in a formula of C_3H_7. Calculating the saturation index we arrive at a value of 0.5

$$R + DB = 3 - 1/2(7) + 1/2(0) + 1 = 0.5$$

and since this value ends in 0.5 we know that it is an EE ion.

Let us compare what we know about this fragment with the possible molecular ions we have already identified. Clearly, it is impossible to have this fragment from either C_3H_3F or C_3H_6O since there are not enough hydrogens in these molecules to give us a fragment with the formula C_3H_7.

We can now conclude that this spectrum must be of a compound with the formula C_4H_{10}. Can we identify the specific compound? In principle, the answer to this question is yes, but in practice the student will find it very difficult. In many cases the spectra of isomers are very similar to each other. How then can we finish the identification procedure?

There are several possibilities. We could compare our unknown spectrum with a spectrum from some standard database such as the NIST Mass Spectral Database. We could compare our unknown spectrum with authentic spectra (spectra of known compounds collected under the same conditions on the same spectrometer). We could take the unknown compound, prepare a chemical derivative of the material and analyze the spectra of the derivatized compound.

It is critically important for the student to recognize the limits of mass spectrometry as well as the strengths. It is beyond the scope of this text to discuss other qualitative analytical methods, but the student should remember that other techniques such as infrared spectroscopy and nuclear magnetic resonance spectrometry can provide a wealth of information about a chemical material. If these other techniques are available then they should be used in conjunction with mass spectrometry to identify unknown materials. In our example, we could be dealing with either butane or methylpropane. A simple proton NMR spectrum of our unknown would allow us to differentiate between these possibilities and conclusively identify our compound.

After determining one or more reasonable formula(e) for our unknown compound, the next step is to determine the structure of the compound. In order to determine the structure, we need to consider the other lines in the mass spectrum and determine what kind of logical losses from the parent compound would produce these lines. How do we go about the task of identifying logical losses?

One reasonable approach to solving a spectrum is to subtract the mass of the fragment peaks from the mass of the parent in order to determine which neutral (either a radical or a small, stable molecule) has been lost from the parent compound. Table 2 (at the end of this chapter) is a tabulation of the common mass losses for these neutrals and the types of compounds indicated by these losses. Although Table 2 is a useful tool, the student is reminded that this list is not exhaustive and that different types of compounds can lose the same kinds of neutrals. You must also remember that different neutrals can have the same mass (28 can be C_2H_4 or CO). Another approach is to tabulate the most common ions observed for particular values of m/z. In this case, some of the lines in a given spectrum can be assigned to certain ions. Table 3 (also at the end of this chapter) is such a tabulation and can be used with the same precautions.

Let us use these two tables to interpret the following spectrum for Unknown No. 3 (Figure 3.1).

Unknown No. 3

The highest mass in the spectrum is m/z 107 and is probably an isotope satellite. This would indicate that m/z 106 is the molecular ion. The base peak (the line with the greatest intensity) is m/z 105. Clearly, this line indicates the loss of hydrogen from m/z 106. Table 2 indicates that this is a common loss for aldehydes, acetals, aryl$-CH_3$ groups, $N-CH_3$ groups, $-CH_2CN$ and alkynes.

For the purposes of this example we will ignore the possibility that nitrogen is present. If m/z 77 is formed from the molecular ion, then a neutral of mass 29 has been lost, indicating an aromatic aldehyde, phenol, ethyl derivative or alkane. What if m/z 77 is

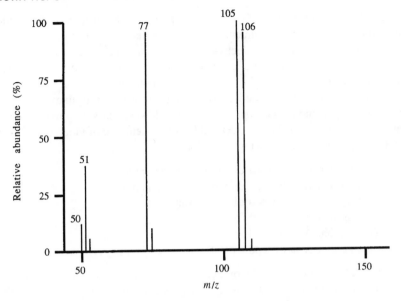

Figure 3.1

formed from 105? The loss of 28 indicates C_2H_4 and CO as common neutral losses. The loss of 28 is consistent with phenols, aldehydes, diaryl ethers, quinones and ethyl esters.

Finally, we look at the ion table (Table 3). For m/z 77, the formula C_6H_5 is common (a benzene ring minus one hydrogen). For m/z 105 we could have $C_6H_5CO^+$, $C_6H_5CH_2CH_2$ or $C_6H_5CHCH_3$.

For the sake of clarity, let us tabulate our observations and deductions:

m/z	Observations/deductions
107	Isotope satellite
106	Molecular ion
105	Loss of H from molecular ion
	Possible compounds: aldehydes, acetals, aryl—CH_3 groups, alkynes
77	Loss of 29 from 106
	Possible compounds: aromatic aldehyde, phenol, ethyl derivative or alkane
	Loss of 28 from 105
	Possible compounds: phenols, aldehydes, diaryl ethers, quinones and ethyl esters

Possible structures or formulae:

77   C_6H_5

105

Clearly the only structure for m/z 105 that fits with our previous observations is $C_6H_5CO^+$, indicating an aromatic aldehyde. The unknown compound is benzaldehyde.

This example was fairly straightforward and may lead the student astray because of its simplicity. The student is cautioned that not all lines in a given spectrum will be formed directly from the molecular ion. In fact, many of the lines present in a very complicated spectrum will be due to further fragmentation of higher mass ions. The general procedure to be used is to start with the molecular ion line and subtract the mass values for corresponding major lines from the molecular ion line. This process is repeated for successive ion lines moving down in mass from the molecular ion line.

As an example, consider a hypothetical spectrum composed of four lines. We will label these lines as M (the molecular ion line), A (the lowest mass fragment line), B (a fragment line of greater mass than line A) and C (a fragment line of greater mass than B). We would then perform the following operations:

M − C   assuming that C is formed directly from the parent.
M − B   assuming that B is formed directly from the parent.
M − A   assuming that A is formed directly from the parent.
C − B   assuming that B is formed from C.
C − A   assuming that A is formed from C.
B − A   assuming that A is formed from B.

We would then consult the table of neutral fragments (Table 2) to determine the possible types of compounds that would be consistent with these losses. Finally, we would consult Table 3 to see what formulae or structures would correspond to the A, B and C ions. From all of the resulting information, we would then propose a logical structure for our unknown. Of course, if we have a very complicated spectrum then we should expect to have many more operations to perform.

Unknown No. 4

Consider the following spectrum (Figure 3.2):

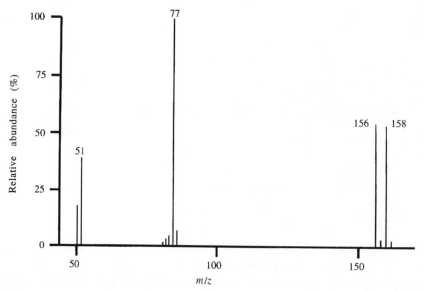

Figure 3.2

m/z	% Relative abundance
50	17.2
51	38.9
74	3.2
75	5.7
76	5.7
77	100.0
78	7.8
156	64.1
157	4.2
158	63.7
159	3.8

The most obvious feature of this spectrum is the characteristic $A+2$ pattern of a monobrominated compound, which is indicated by the lines at m/z 156 and 158. Inserting our error limits and scaling the 158 line to 100% gives us:

m/z	% Relative abundance
158	100.0
159	6.0 ± 0.6

which indicates either five or six carbons. Our two possible starting formulae are:

(1) C_5Br

(2) C_6Br

and there are no other important $A+2$ elements with the possible exception of oxygen. Which other elements could be present?

1. Subtracting 141 (the combined mass of five carbons and one ^{81}Br) from 158 leaves 17 mass units. Notice that in calculating the mass of C_5Br we used 81 for the bromine, since this would be the ^{81}Br isotope. It is important for the student to keep in mind the distinct isotopic forms of the elements when determining molecular formulae.

From the residual mass, we could either have one oxygen and one hydrogen for a formula of C_5HOBr, or we could have 17 hydrogens for a formula of $C_5H_{17}Br$. The latter would be impossible since it violates the $2n+2$ rule which tells us that for five carbons we can have 12 hydrogens. The former gives us a saturation index of 5, indicating a very unsaturated molecule. This would seem to be very unlikely.

2. Subtracting 153 (the combined mass of six carbons and one ^{81}Br) from 158 leaves five mass units. The only possibility is five hydrogens, giving us a formula of C_6H_5Br. The saturation index for this compound is 4 and it meets the requirements of the OE test.

The base peak at m/z 77 indicates the loss of 79 from m/z 156 (or the loss of 81 from m/z 158). From Table 2, this would indicate the loss of Br, which is definitely a logical loss since we know that bromine is present in the compound. Consulting Table 3 for likely formulae for m/z 77, we see that C_6H_5 is likely and we can identify Unknown No. 4 as bromobenzene.

Practice problems

Each of the following problems involves the determination of reasonable molecular formulae and saturation indices. For convenience, the data are presented in tabular format indicating the m/z value and the normalized abundance for the $A+1$ and $A+2$ lines. For this exercise, the only elements present in these compounds are C, H, N, and O.

Based upon the data presented:

1. Determine all of the reasonable molecular formulae using the above elements.
2. For each formula, calculate the saturation index.
3. Based upon the saturation index, determine whether or not the molecular formula represents a possible molecular ion.

Detailed solutions to these problems are available in the Appendix.

Problem 1

m/z	Normalized abundance
44	100.0%
45	1.2%
46	0.4%

Problem 2

m/z	Normalized abundance
30	100.0%
31	2.3%
32	0.0%

Problem 3

m/z	Normalized abundance
84	100.0%
85	6.7%
86	0.0%

Problem 4

m/z	Normalized abundance
56	100.0%
57	4.7%
58	0.1%

Problem 5

m/z	Normalized abundance
91	100.0%
92	7.6%
93	0.3%

Problem 6

m/z	Normalized abundance
46	100.0%
47	2.9%
48	0.2%

Problem 7

m/z	Normalized abundance
106	100.0%
107	7.6%
108	0.4%

Problem 8

m/z	Normalized abundance
93	100.0%
94	7.1%
95	0.2%

Table 2 Mass and formulae of common neutral particles. Note that neutrals can be lost as molecules or as radicals. In some cases, the formulae of neutral particles will be the same as for fragment ions (Table 3). This list is not meant to be exhaustive

Mass lost	Neutral formula	Compound types indicated
1	H	Aldehydes, acetals, compounds with aryl$-CH_3$ groups, compounds with N-CH_3, compounds with $-CH_2CN$, alkynes.
2	H_2	Fused ring aromatic compounds.
14	N	Aryl$-NO$.
15	CH_3	Acetals, methyl derivatives, t-butyl and i-propyl compounds, compounds with aryl$-C_2H_5$ groups, $(CH_3)_3SiO$ derivatives, alkanes.
16	CH_4, NH_2, O	Aromatic nitro compounds, N-oxides, S-oxides, aromatic amides.
17	NH_3, OH	Carboxylic acids, aromatic compounds with a functional group containing oxygen ortho to one containing hydrogen, e.g. o-nitrotoluene, amines.
18	H_2O	Alcohols, aldehydes, ketones, ethers, carboxylic acids.
19	F	Fluoroalkanes.
20	HF	Fluoroalkanes.
26	C_2H_2	Aromatic hydrocarbons.
27	C_2H_3, HCN	Aromatic amines, aromatic nitriles, nitrogen heterocycles.
28	C_2H_4, N_2, CO	Phenols, aldehydes, diaryl ethers, quinones, aliphatic nitriles, ethyl esters.
29	CH_3N, C_2H_5 CHO	Aromatic aldehydes, phenols, aliphatic nitriles, ethyl derivatives, alkanes.

(*continued*)

Table 2 (*continued*)

Mass lost	Neutral formula	Compound types indicated
30	C_2H_6, NO, CH_2O NH_2CH_2	Aromatic nitro compounds, aromatic methyl ethers.
31	CH_3O	Methoxy derivatives, methyl esters.
32	CH_3OH, S	Ethers, carboxylic acids, sulfur-containing compounds.
33	$CH_3 + H_2O$, CH_2F	Alcohols, fluoroalkanes.
34	H_2S	Thiols
35 (and 37)	Cl	Chloroalkanes
36	HCl	Chloroalkanes
38	H_2O_2	Polycarboxylic acids
40	C_3H_4, CH_2CN	Aliphatic nitriles, aromatics.
41	C_3H_5	Propyl esters
42	C_3H_6, CH_2CO	Acetates, *N*-acetyl compounds, butyl ketones.
43	HNCO, CH_3CO, C_3H_7	Propyl derivatives, aliphatic nitriles, amides, methyl ketones, alkanes.
44	C_3H_8, $CONH_2$, CH_2CO, CO_2, CS	Aldehydes, esters, acids, amides, thiophenols, aryl-S-aryl.
45	C_2H_5O	Carboxylic acids, ethyl esters, ethers.
46	$[H_2O + C_2H_4]$, NO_2, CH_2S	Methyl aryl sulfides, cyclothioalkanes, nitro-aromatics, nitroesters, ethers, carboxylic acids.
47	C_2H_4F, CH_3S	Fluoroalkanes, thiols.
49	CH_2Cl	Chloroalkanes.
51	C_3HN	Nitrogen heterocyclics.
54	C_4H_6, C_3H_2O	Aromatics, cyclic-CO−CH=CH−
55	C_4H_7	Butyl esters.
56	CH_2=$CHCH_2CH_3$, CH_3CH=$CHCH_3$, 2CO	Pentyl ketones, ArOBu ethers.
57	C_4H_9, C_2H_5CO	Alkanes, ethers, ketones.
58	C_2H_6CO	Ketones
59	C_3H_7O, CH_3CO_2	Ethers, esters.
60	COS, CH_3COOH, C_3H_7OH	Ethers, esters, thiocarbamates.
61	C_2H_5S	Thiols, thioethers.
64	SO_2, $(CH_3OH)_2$	Dialcohols, RSO_2R, Aryl−SO_2OR
79 (and 81)	Br	Bromo compounds
127	I	Iodo compounds

Table 3 Mass and formulae for common ion fragments. In some cases, fragment ion formulae will be the same as for neutral particles (Table 2). This list is not meant to be exhaustive

Mass	Fragment ion formulae (all ions are positively charged)
15	CH_3^+
16	O^+
17	OH^+
18	H_2O^+, NH_4^+
19	F^+, H_3O^+
26	CN^+
27	$C_2H_3^+$
29	$C_2H_5^+$, CHO^+
30	$CH_2NH_2^+$, NO^+
31	CH_2OH^+, OCH_3^+
32	O_2^+

(*continued*)

Table 3 (*continued*)

Mass	Fragment ion formulae (all ions are positively charged)
34	H_2S^+
35 (and 37)	Cl^+
36	HCl^+
39	$C_3H_3^+$
43	$C_3H_7^+$, $CH_3C{=}O^+$, $C_2H_5N^+$
44	$[CH_2CHO + H]^+$, $CH_3CHNH_2^+$, CO_2^+, $NH_2CC{=}O^+$, $(CH_3)_2N^+$
45	CH_3CHOH^+, $CH_2CH_2OH^+$, $CH_2OCH_3^+$, $COOH^+$
47	CH_2SH^+, CH_3S^+
49	CH_2Cl^+
53	$C_4H_5^+$
55	$C_4H_7^+$, $CH_2{=}CHC{=}O^+$
56	$C_4H_8^+$
57	$C_4H_9^+$, $C_2H_5C{=}O^+$
59	$(CH_3)_2COH^+$, $CH_2OC_2H_5^+$, $COOCH_3^+$, $CH_3OCHCH_3^+$, $CH_3CHCH_2OH^+$

Mass	
65	
69	$C_5H_9^+$, CF_3^+, $CH_3CH{=}CHC{=}O^+$, $CH_2{=}C(CH_3)C{=}O^+$
77	(C_6H_5)
91	(C_7H_7)
92	
93	CH_2Br^+, $C_7H_9^+$ (terpenes)
105	
107	

4 Hydrocarbons

We will begin this chapter by looking at three general classes of alkanes; linear, branched and cyclic. The spectra of ordinary hydrocarbons provide a convenient starting point for learning general interpretation procedures.

Linear alkanes

Example 1   Methane (CH_4)

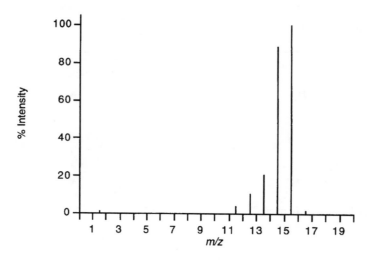

m/z	%
2	1.2
1 2	3.8
1 3	10.7
1 4	20.4
1 5	88.8
1 6	100.0
1 7	1.6

Example 2   Ethane (C$_2$H$_6$)

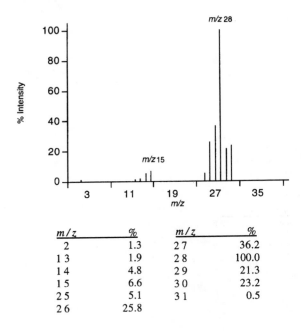

m/z	%	m/z	%
2	1.3	27	36.2
13	1.9	28	100.0
14	4.8	29	21.3
15	6.6	30	23.2
25	5.1	31	0.5
26	25.8		

Example 3   Butane (C$_4$H$_{10}$)

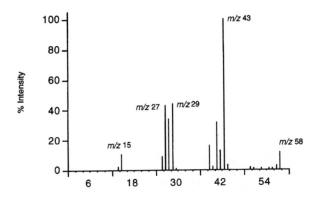

m/z	%	m/z	%
14	2.0	42	12.6
15	10.8	43	100.0
26	9.0	44	3.4
27	42.8	50	1.9
28	33.8	51	1.4
29	44.2	53	1.0
30	1.0	55	1.2
39	17.3	57	2.7
40	2.4	58	11.8
41	31.8	59	0.5

Example 4   Hexane (C$_6$H$_{14}$)

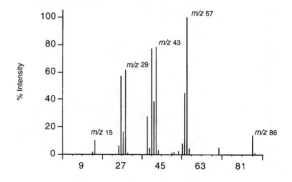

m/z	%
1 4	1.4
1 5	10.2
2 6	6.4
2 7	56.9
2 8	16.1
2 9	61.2
3 0	1.3
3 9	27.3
4 0	4.2
4 2	38.8
4 3	78.0

m/z	%
4 4	2.6
5 0	1.3
5 1	1.8
5 3	2.5
5 5	8.0
5 6	44.8
5 7	100.0
5 8	4.5
7 1	5.2
8 6	14.0
8 7	0.9

Example 5   Octane (C$_8$H$_{18}$)

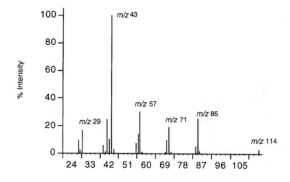

m/z	%
2 7	9.8
2 8	2.2
2 9	16.9
3 9	5.5
4 0	1.1
4 1	24.5
4 2	10.0
4 3	100.0
4 4	2.9
5 5	7.1
5 6	14.0

m/z	%
5 7	30.3
5 8	1.4
6 9	1.0
7 0	9.6
7 1	19.2
7 2	1.0
8 4	4.8
8 5	25.1
8 6	1.6
114	3.0
115	0.2

Example 6   Dodecane (C$_{12}$H$_{26}$)

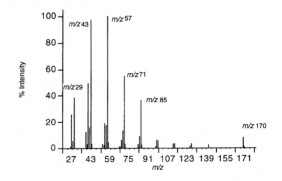

m/z	%	m/z	%	m/z	%
26	1.3	57	100.0	86	2.4
27	25.4	58	4.5	97	1.1
28	5.7	67	1.3	98	6.1
29	38.4	68	1.0	99	5.8
39	12.2	69	6.4	112	3.6
40	49.3	70	13.3	113	3.4
43	15.5	71	54.5	126	1.7
44	3.2	72	3.1	127	3.6
53	2.9	83	2.8	141	2.1
54	2.1	84	8.9	170	7.9
55	18.9	85	36.3	171	1.1
56	17.6				

Example 7   Tetradecane (C$_{14}$H$_{30}$)

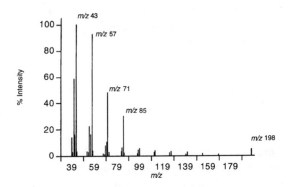

m/z	%	m/z	%	m/z	%
39	14.1	67	1.8	98	4.4
40	2.9	68	1.4	99	5.8
41	58.4	69	7.6	112	3.0
42	16.0	70	10.7	113	3.7
43	100.0	71	48.0	126	2.1
44	3.4	72	2.8	127	3.2
53	3.2	83	3.3	140	1.4
54	2.7	84	5.9	141	2.6
55	22.6	85	30.4	155	1.7
56	16.4	86	2.1	169	1.0
57	92.6	97	1.8	198	5.0
58	4.1				

Ordinary alkanes are characterized as having many relatively weak C−C bonds. The spectra of saturated alkanes normally exhibit many lines, with the lines occurring as distinct recognizable groups. In any given group, there will typically be a fragment of the formula C_nH_{2n+1} (an alkyl ion), the ^{13}C isotope peak and additional peaks corresponding to C_nH_{2n}, C_nH_{2n-1} and C_nH_{2n-2}. In any given group, the alkyl peak will tend to have the greatest intensity.

In general, the lower mass fragments corresponding to two, three, four and perhaps five carbons will have greater intensity than fragments containing six or more carbons. Two factors affect this difference in intensity. The first factor is that most mass spectrometers tend to discriminate against higher mass fragments. Fewer of the high mass ions are collected and therefore the intensity is lower. Second, the lower mass fragments are typically produced by higher mass fragments undergoing secondary fragmentation or rearrangements. Students are encouraged to study Examples 1–7, paying close attention to the patterns produced by these linear alkanes.

The most common fragmentation mechanism encountered with the linear alkanes is simple cleavage of a C−C single bond to produce a carbocation and an alkyl radical. This can be represented as

$$[CH_3-CH_3]^{\ddot{+}} \longrightarrow CH_3^{\oplus} + CH_3^{\bullet}$$

The carbocation formed will follow the order of abundance $CH_3^+ < RCH_2^+ < R_2CH^+ < R_3C^+$. Of course, with a linear alkane there is no possibility of producing a tertiary carbocation unless significant rearrangement occurs, but the migration of hydrogen from an adjacent carbon can result in the formation of a secondary carbocation.

A secondary mechanism involves the loss of an olefin molecule

$$R-CH_2-CH_2-CH_2^{\oplus} \longrightarrow R-CH_2^{\oplus} + CH_2=CH_2$$

resulting in a new ion 28 mass units smaller. This mechanism accounts for the typical pattern shown by most alkanes: clusters of fragments separated by 14 amu. Please pay careful attention to the arrows shown in this and all other examples. A double headed arrow is used to show the movement of a pair of electrons. Single headed arrows are used to show the movement of a single electron.

In general, linear alkanes will show significant intensity for the molecular ion peak. As the amount of branching increases, the intensity of the molecular ion peak will decrease, and when the molecule has significant branching the molecular ion may be undetectable.

While the overall appearance of an ordinary alkane spectrum can be characterized as a series of fragments differing by 14 amu, the student should be aware that the loss of 14 amu is not generally considered to be a logical loss. What accounts for the apparent loss of 14 amu? Consider the spectrum of dodecane (Example 6).

In this spectrum, we can have the initial fragmentation occur between any two adjacent carbons. If we initially lose a methyl radical (due to fragmentation between carbons 1 and 2) followed by successive loss of ethene (a neutral olefin molecule) the resulting ion would have a mass of 127. This fragment and other smaller fragment ions can also lose ethene, and we can easily imagine m/z 127 losing 28 to form m/z 99, which loses 28 to form m/z 71, which loses 28 to form m/z 43. Similarly, we can imagine an initial fragmentation occurring between carbons 2 and 3 producing an ethyl radical and a fragment ion with a mass of 141. Loss of ethene from this fragment produces an ion with a mass of 113, and the ion at m/z 113 loses 28 to form m/z 85, which loses 28 to form m/z 57. When we add these two patterns together we are left with the impression that a single hydrocarbon chain is being broken down, one CH_2 unit at a time.

Other mechanisms such as the loss of hydrogen do occur but are generally not important, other than adding to the complexity of the spectra.

Branched alkanes

Example 8   2-Methylpropane (C_4H_{10})

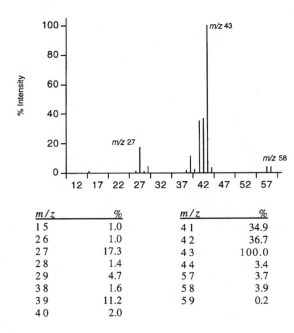

m/z	%	m/z	%
1 5	1.0	4 1	34.9
2 6	1.0	4 2	36.7
2 7	17.3	4 3	100.0
2 8	1.4	4 4	3.4
2 9	4.7	5 7	3.7
3 8	1.6	5 8	3.9
3 9	11.2	5 9	0.2
4 0	2.0		

Example 9   2-Methylpentane (C$_6$H$_{14}$)

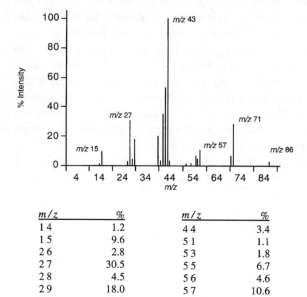

m/z	%		m/z	%
1 4	1.2		4 4	3.4
1 5	9.6		5 1	1.1
2 6	2.8		5 3	1.8
2 7	30.5		5 5	6.7
2 8	4.5		5 6	4.6
2 9	18.0		5 7	10.6
3 9	20.0		7 0	6.8
4 0	3.2		7 1	28.5
4 1	35.5		8 6	2.9
4 2	52.8		8 7	0.2
4 3	100.0			

Example 10   2,2-Dimethylhexane (C$_8$H$_{18}$)

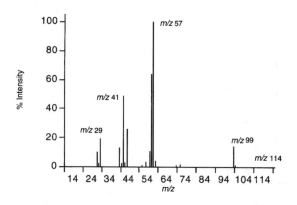

m/z	%		m/z	%
2 7	10.2		5 3	3.3
2 8	2.0		5 5	10.7
2 9	19.2		5 6	63.5
3 9	12.7		5 7	100.0
4 0	2.1		5 8	4.2
4 1	48.5		6 9	1.4
4 2	2.8		7 1	1.6
4 3	25.7		9 9	13.9
5 1	1.2		100	1.0

Example 11   2,2,3,3,5,6,6-Heptamethylheptane ($C_{14}H_{30}$)

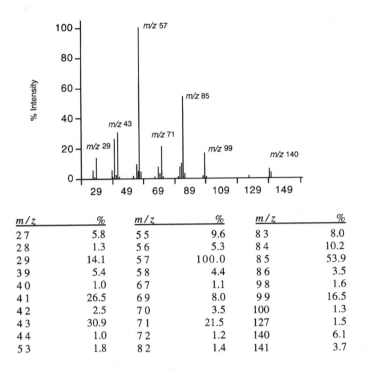

m/z	%	m/z	%	m/z	%
2 7	5.8	5 5	9.6	8 3	8.0
2 8	1.3	5 6	5.3	8 4	10.2
2 9	14.1	5 7	100.0	8 5	53.9
3 9	5.4	5 8	4.4	8 6	3.5
4 0	1.0	6 7	1.1	9 8	1.6
4 1	26.5	6 9	8.0	9 9	16.5
4 2	2.5	7 0	3.5	100	1.3
4 3	30.9	7 1	21.5	127	1.5
4 4	1.0	7 2	1.2	140	6.1
5 3	1.8	8 2	1.4	141	3.7

The spectrum of a branched alkane is significantly different from that of the corresponding linear alkane. Fragmentation at the branching points in the molecule allows the formation of secondary or tertiary carbocations, which are much more stable than primary carbocations. This enhanced stability results in greater intensity for these lines in the mass spectrum.

Compare the spectrum of Example 8 (2-methylpropane or isobutane) with Example 3 (butane). The principal differences between these two spectra are a decrease in the intensity of the parent ion in the 2-methylpropane and the almost complete disappearance of m/z 29 corresponding to the ethyl cation. The formation of this ethyl cation in 2-methylpropane would require significant rearrangement, whereas in the butane molecule simple fragmentation with allow the formation with relative ease.

Example 9 (2-methylpentane) compared with Example 4 (hexane) shows an enhancement of m/z 71 due to loss of methyl from either position in the 2-methylpentane. The fragment at m/z 57 is virtually absent owing to the difficulty of producing $C_4H_9{}^+$, while m/z 43 is enhanced as there are two different $C_3H_7{}^+$ ions which can be formed by cleavage of the bond between the second and the third carbon in the chain. These possibilities are shown below (Figure 4.1).

Example 10 (2,2-dimethylhexane, Figure 4.2) shows a significant peak at m/z 99 compared with Example 5 (octane), where there is no such fragment. This is because, in the 2,2-dimethylhexane, any of three methyls may be lost, resulting in a very stable tertiary carbocation. Furthermore, the greatly enhanced m/z 57 is due to cleavage between

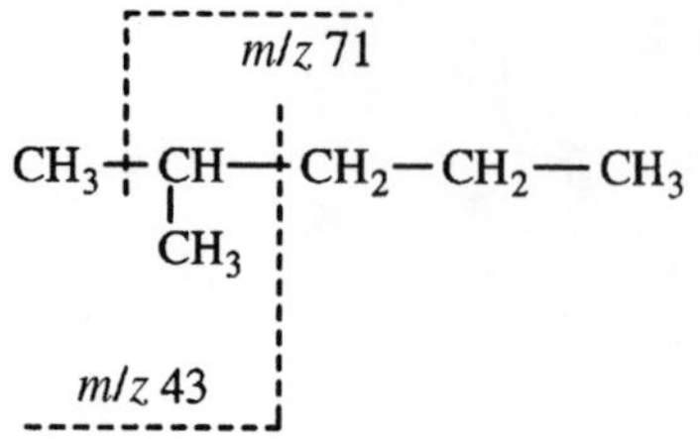

Figure 4.1

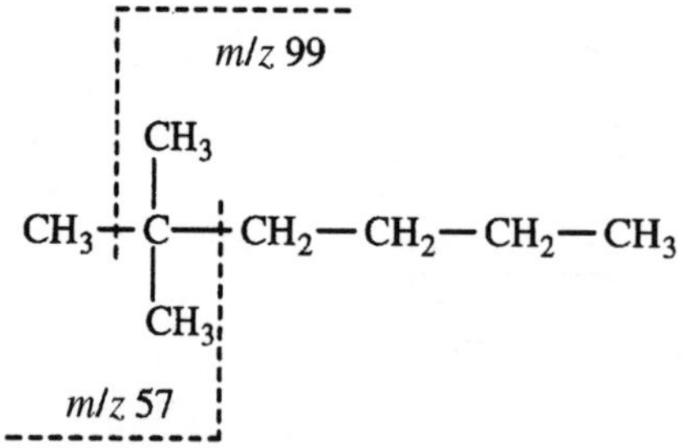

Figure 4.2

the second and third carbons resulting in two possible $C_4H_9^+$ fragments, one of which is also a tertiary carbocation.

Finally, Example 11 (2,2,3,3,5,6,6-heptamethylheptane, Figure 4.3) shows a significantly reduced m/z 71 compared with Example 7 (tetradecane) owing to the greater difficulty in forming $C_5H_{11}^+$. The abundance of m/z 85 is enhanced since it is possible to fragment the molecule between carbons 4 and 5, producing a secondary carbocation. It is possible to form two fragments corresponding to $C_7H_{15}^+$ (m/z 99) by breaking the bond between the number 3 and 4 carbons, and one of these fragments would be a very stable tertiary carbocation. Likewise, there are two possible $C_4H_9^+$ fragments (m/z 57), each of which is the more stable tertiary carbocation.

In general, the student will find that the spectra of branched alkanes are governed by the loss of the largest possible alkyl radical (R) and the formation of the most stable carbocation. This is an example of Stevenson's rule, which tells us that the largest alkyl radical will be lost preferentially.

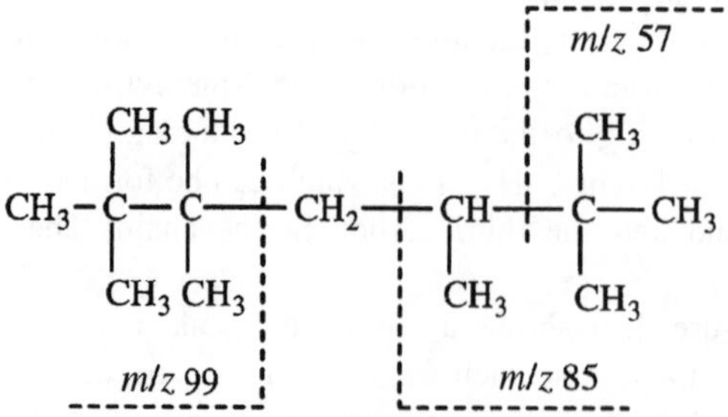

Figure 4.3

Cyclic alkanes

Example 12   Cyclohexane (C_6H_{12})

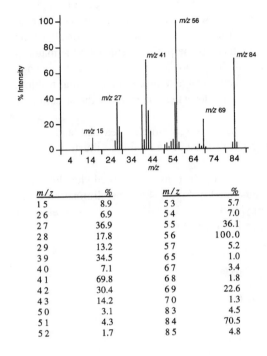

m/z	%	m/z	%
15	8.9	53	5.7
26	6.9	54	7.0
27	36.9	55	36.1
28	17.8	56	100.0
29	13.2	57	5.2
39	34.5	65	1.0
40	7.1	67	3.4
41	69.8	68	1.8
42	30.4	69	22.6
43	14.2	70	1.3
50	3.1	83	4.5
51	4.3	84	70.5
52	1.7	85	4.8

Example 13   Methylcyclohexane (C_7H_{14})

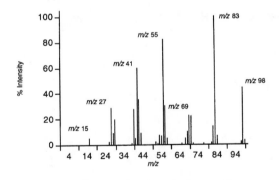

m/z	%	m/z	%	m/z	%
14	4.9	51	2.1	70	21.6
26	2.0	52	1.1	71	1.4
27	28.7	53	7.1	77	1.0
28	8.8	54	6.5	81	1.5
29	19.5	55	82.1	82	14.2
38	1.1	56	29.9	83	100.0
39	27.4	57	4.9	84	6.5
40	5.0	65	1.1	97	2.3
41	59.8	67	5.0	98	44.0
42	35.4	68	9.9	99	3.3
43	8.8	69	22.6		

In general the molecular ions formed from cycloalkanes are more abundant than those formed from their acyclic analogs (see m/z 84 in Example 12 and m/z 98 in Example 13). The interpretation of the structures of these cycloalkanes is also much more difficult than for the corresponding non-cyclic forms. The decomposition of the carbon skeleton typically involves cleavage at two bond positions, which can contribute to considerable randomization. The elimination of an olefin from the molecular ion will result in the formation of additional odd-electron ions which can be helpful in identifying the material as a cycloalkane. When the ring structure of the cycloalkane opens, simple cleavage reactions similar to those seen for the linear alkanes can take place in the normal fashion. Thus, there are two important mechanisms to consider when interpreting the structure of a cycloalkane.

Formation of the even-electron ion (m/z 69) through simple cleavage and of the important odd-electron ion (m/z 56) through elimination of an olefin are illustrated in the following mechanism (Figure 4.4).

Fragmentations producing the two important odd-electron ions in methylcyclohexane plus the even-electron ion base peak are illustrated in Figure 4.5. The elimination of an olefin from the open ring structure in methylcyclohexane accounts for the odd-electron peak at m/z 70. This reaction is summarized in Figure 4.6.

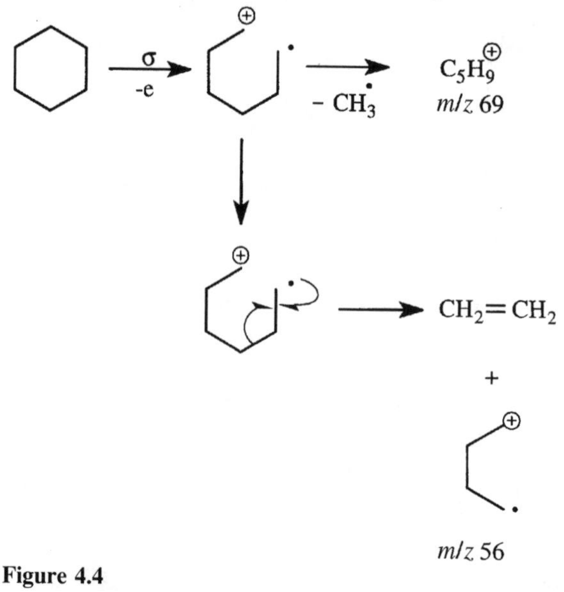

Figure 4.4

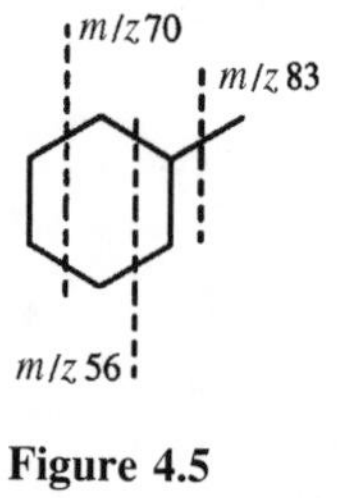

Figure 4.5

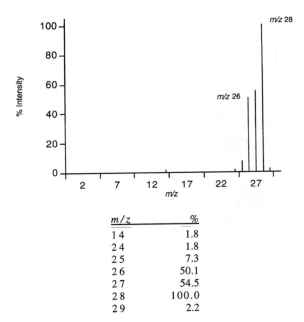

m/z 70

Figure 4.6

Note that in the above example we have chosen to place the positive charge on the most highly substituted carbon. This is the usual and customary procedure and the student should follow this procedure whenever possible.

Alkenes

Example 14   Ethene (C_2H_4)

m/z	$\%$
1 4	1.8
2 4	1.8
2 5	7.3
2 6	50.1
2 7	54.5
2 8	100.0
2 9	2.2

Example 15   Propene (C₃H₆)

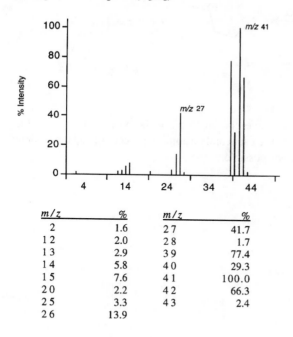

m/z	%	m/z	%
2	1.6	27	41.7
12	2.0	28	1.7
13	2.9	39	77.4
14	5.8	40	29.3
15	7.6	41	100.0
20	2.2	42	66.3
25	3.3	43	2.4
26	13.9		

Example 16   (a) 1-Butene (C₄H₈)

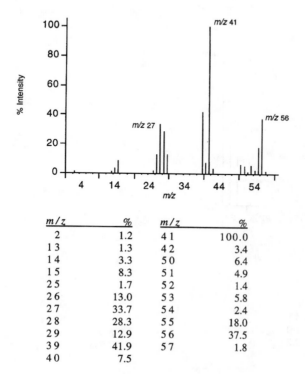

m/z	%	m/z	%
2	1.2	41	100.0
13	1.3	42	3.4
14	3.3	50	6.4
15	8.3	51	4.9
25	1.7	52	1.4
26	13.0	53	5.8
27	33.7	54	2.4
28	28.3	55	18.0
29	12.9	56	37.5
39	41.9	57	1.8
40	7.5		

Example 16   (b) 2-Butene

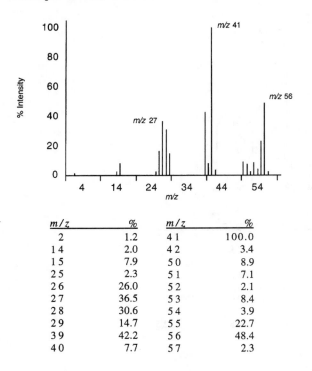

m/z	%	m/z	%
2	1.2	4 1	100.0
1 4	2.0	4 2	3.4
1 5	7.9	5 0	8.9
2 5	2.3	5 1	7.1
2 6	26.0	5 2	2.1
2 7	36.5	5 3	8.4
2 8	30.6	5 4	3.9
2 9	14.7	5 5	22.7
3 9	42.2	5 6	48.4
4 0	7.7	5 7	2.3

Example 17   (a) 1-Hexene (C_6H_{12})

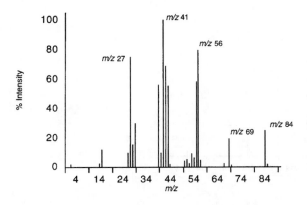

m/z	%	m/z	%	m/z	%
2	1.6	4 1	100.0	5 5	58.3
1 4	2.4	4 2	68.5	5 6	79.2
1 5	11.6	4 3	55.3	5 7	4.2
2 6	9.6	4 4	1.9	6 7	2.1
2 7	74.7	5 0	3.9	6 9	19.2
2 8	15.1	5 1	5.1	7 0	1.2
2 9	29.7	5 2	2.0	8 4	24.6
3 9	56.0	5 3	8.8	8 5	1.7
4 0	9.6	5 4	6.3		

Example 17   (b) 3-Hexene

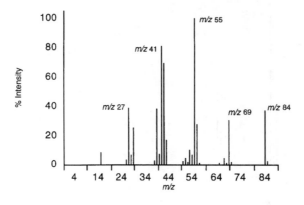

m/z	%	m/z	%	m/z	%
1 5	8.5	4 2	69.3	5 7	1.2
2 6	3.3	4 3	16.5	6 5	1.4
2 7	38.5	5 0	2.1	6 7	4.7
2 8	6.9	5 1	4.3	6 8	1.1
2 9	25.3	5 2	1.9	6 9	30.3
3 8	3.0	5 3	10.2	7 0	1.6
3 9	38.1	5 4	6.9	8 4	36.9
4 0	7.4	5 5	100.0	8 5	2.5
4 1	81.2	5 6	27.6		

Example 18   (a) 1-Octene (C_8H_{16})

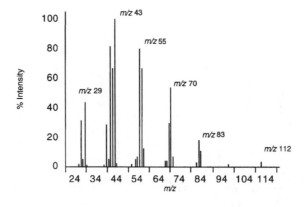

m/z	%	m/z	%	m/z	%
2 6	1.8	4 3	100.0	6 9	29.6
2 7	31.3	4 4	2.2	7 0	53.8
2 8	5.0	5 1	1.6	7 1	6.7
2 9	43.7	5 3	5.3	8 2	2.8
3 0	1.1	5 4	6.4	8 3	17.8
3 8	1.2	5 5	79.9	8 4	10.7
3 9	28.5	5 6	66.6	9 7	1.5
4 0	4.8	5 7	12.6	112	3.6
4 1	81.8	6 7	3.8	113	0.2
4 2	66.5	6 8	4.1		

Example 18   (b) 4-Octene

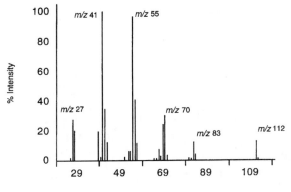

m/z	%	m/z	%	m/z	%
26	1.9	53	6.3	69	24.2
27	27.6	54	6.4	70	30.4
29	20.3	55	96.4	71	3.4
39	19.6	56	40.7	81	1.7
40	2.1	57	12.0	82	1.1
41	100.0	65	1.3	83	12.5
42	34.8	66	1.1	84	3.8
43	12.2	67	7.1	112	12.8
51	2.3	68	2.7	113	1.1

Example 19   2-Methyl-1-propene (C_4H_8)

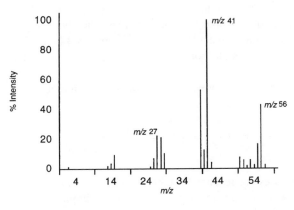

m/z	%	m/z	%
2	1.1	41	100.0
13	1.4	42	3.7
14	3.5	50	7.1
15	9.0	51	5.5
26	6.7	52	1.5
27	21.9	53	5.5
28	20.4	54	2.3
29	10.2	55	16.4
39	52.7	56	42.3
40	12.0	57	2.1

Example 20   Cyclobutane (C$_4$H$_8$)

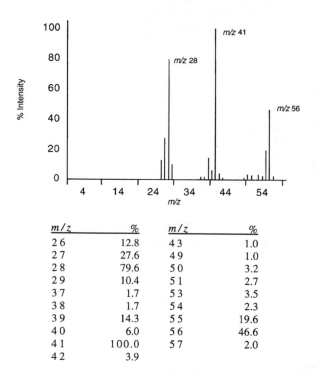

m/z	%	m/z	%
2 6	12.8	4 3	1.0
2 7	27.6	4 9	1.0
2 8	79.6	5 0	3.2
2 9	10.4	5 1	2.7
3 7	1.7	5 3	3.5
3 8	1.7	5 4	2.3
3 9	14.3	5 5	19.6
4 0	6.0	5 6	46.6
4 1	100.0	5 7	2.0
4 2	3.9		

The presence of a double bond increases the abundance of the $C_nH_{2n-1}{}^+$ and $C_nH_{2n}{}^+$ series of ions. In general, the presence of a double bond will increase the abundance of the molecular ion for compounds with low molecular weights (Examples 14–18). The normal ionization site in an alkene is the double bond, as opposed to ionization at any C–C bond in the alkanes.

Alkenes tend to undergo allylic cleavage as shown in the following example:

$$CH_3 - CH_2 - CH \overset{\bullet}{-} \overset{\oplus}{CH} - R \longrightarrow CH_3\bullet + CH_2 = CH \overset{\oplus}{-} CH - R$$

The ions produced by these processes are resonance stabilized (Figure 4.7).

In propene (Example 15) the peak at m/z 27 is consistent with simple cleavage resulting in the loss of the methyl radical. This is also seen in 1-butene (Example 16a). The base peak at m/z 41 can be formed either through a simple cleavage mechanism between the number 3 and 4 carbons or through the allylic cleavage mechanism shown above. Preference would be given to the simple cleavage mechanism, since the spectrum of 2-butene (Example 16b) is virtually identical with that of 1-butene. In 2-butene, the double bond would have to migrate before the allylic cleavage mechanism would be possible.

The double bond in the molecular ion is very mobile and as a consequence the spectra of isomeric alkenes are very similar unless the double bond is in close proximity to a

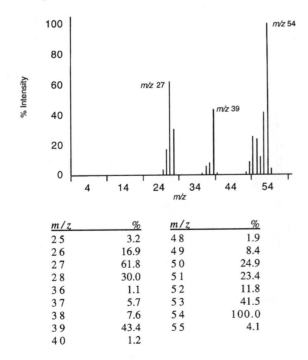

Figure 4.7

functional group of some sort. Students are asked to compare Examples 16a, 16b, and 19; 17a and b; and 18a and b. Cycloalkanes, will also give spectra very similar to those of alkenes (compare Example 20, cyclobutane with 16a and b and 19 or Example 12, cyclohexane with 17a and b).

Dienes

Example 21   (a) 1,2-Butadiene (C_4H_6)

m/z	%	m/z	%
2 5	3.2	4 8	1.9
2 6	16.9	4 9	8.4
2 7	61.8	5 0	24.9
2 8	30.0	5 1	23.4
3 6	1.1	5 2	11.8
3 7	5.7	5 3	41.5
3 8	7.6	5 4	100.0
3 9	43.4	5 5	4.1
4 0	1.2		

Example 21   (b) 1,3-Butadiene

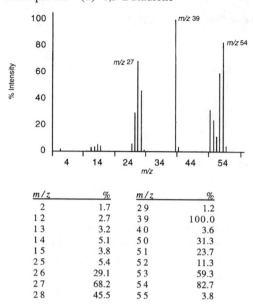

m/z	%	m/z	%
2	1.7	29	1.2
12	2.7	39	100.0
13	3.2	40	3.6
14	5.1	50	31.3
15	3.8	51	23.7
25	5.4	52	11.3
26	29.1	53	59.3
27	68.2	54	82.7
28	45.5	55	3.8

Not surprisingly, the spectra of isomers containing more than one double bond are also very similar. The presence of the large peak at *m/z* 39 in 1,3-butadiene must result from bond migration prior to loss of a methyl group. As we have seen, the exact location of the double bond can be very difficult to determine unless other structural features are present in the molecule.

Alkynes

Example 22 1-Propyne (C_3H_4)

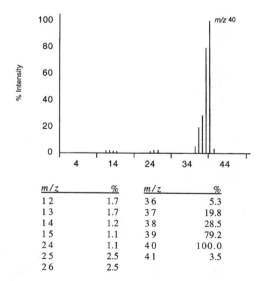

m/z	%	m/z	%
12	1.7	36	5.3
13	1.7	37	19.8
14	1.2	38	28.5
15	1.1	39	79.2
24	1.1	40	100.0
25	2.5	41	3.5
26	2.5		

Example 23   (a) 1-Butyne (C$_4$H$_6$)

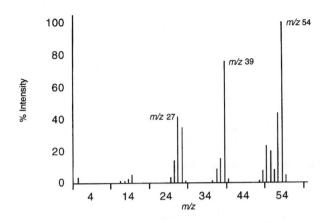

m / z	%	m / z	%	m / z	%
1	3.1	2 8	34.6	4 9	7.5
1 2	1.1	2 9	1.2	5 0	23.0
1 3	1.3	3 6	1.1	5 1	19.7
1 4	2.2	3 7	8.3	5 2	8.0
1 5	5.1	3 8	15.0	5 3	43.3
2 5	3.1	3 9	76.2	5 4	100.0
2 6	14.2	4 0	2.5	5 5	4.3
2 7	41.5	4 8	1.3		

Example 23   (b) 2-Butyne

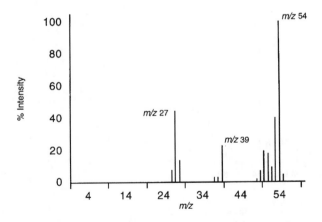

m / z	%	m / z	%
2 6	7.4	4 9	6.7
2 7	44.3	5 0	19.0
2 8	13.6	5 1	17.6
3 7	2.8	5 2	9.0
3 8	2.6	5 3	39.6
3 9	22.4	5 4	100.0
4 8	1.4	5 5	4.2

Example 24   (a) 1-Pentyne (C_5H_8)

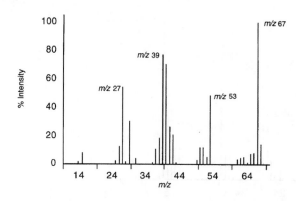

m/z	%	m/z	%	m/z	%
14	1.9	38	18.6	61	3.6
15	8.0	39	77.3	62	4.6
25	2.1	40	70.2	63	5.0
26	12.4	41	26.2	64	1.2
27	54.4	42	21.0	65	7.4
28	2.0	49	2.8	66	8.1
29	29.9	50	11.6	67	100.0
31	3.7	51	12.0	68	14.2
36	1.3	52	4.8	69	0.8
37	10.5	53	48.5		

Example 24   (b) 2-Pentyne

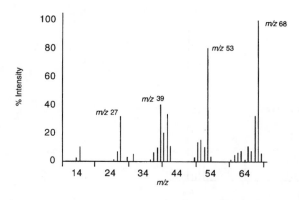

m/z	%	m/z	%	m/z	%
14	2.0	39	40.4	60	1.1
15	10.4	40	20.3	61	4.4
25	1.1	41	33.8	62	6.2
26	6.6	42	10.6	63	7.4
27	32.1	49	3.1	64	1.2
29	3.0	50	13.4	65	10.4
31	4.9	51	15.3	66	7.2
36	1.1	52	10.0	67	32.2
37	6.4	53	80.2	68	100.0
38	9.3	54	3.6	69	5.5

Reactions similar to those examined for the alkenes are also important for the alkynes. Once again, the spectra of isomers are very similar to each other.

In general, determining the exact structure of a given alkene, alkyne or diene is a considerable challenge for the analyst. Normally, one of three methods would routinely be used:

1. Comparison of the unknown spectrum with spectra from a standard database collection (such as the NIST database).
2. Comparison of the unknown spectrum with an authentic spectrum acquired under the same analytical conditions as the unknown.
3. Derivatization of the alkene (or alkyne or diene), followed by determination of the structure of the derivative. Once the structure of the derivative is known, knowledge of the derivatization procedure used and the products formed should result in the identification of the starting material (the original alkene).

Aromatic hydrocarbons

Example 25   Benzene (C_6H_6)

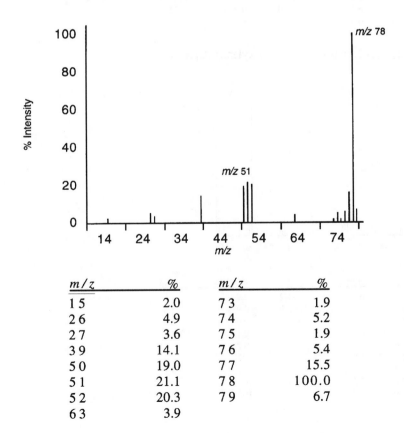

m/z	%	m/z	%
15	2.0	73	1.9
26	4.9	74	5.2
27	3.6	75	1.9
39	14.1	76	5.4
50	19.0	77	15.5
51	21.1	78	100.0
52	20.3	79	6.7
63	3.9		

Example 26   Toluene (C$_7$H$_8$)

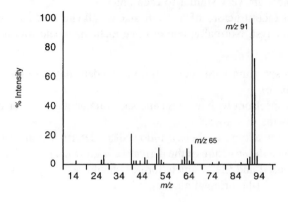

m/z	%	m/z	%	m/z	%
15	2.4	50	7.5	66	1.8
26	3.1	51	10.9	74	1.4
27	6.4	52	2.6	77	1.4
39	20.4	53	1.3	86	1.0
40	2.4	61	2.6	89	4.0
41	2.4	62	5.0	90	5.2
43	2.2	63	10.5	91	100.0
45	4.4	64	2.4	92	72.6
46	3.0	65	13.6	93	5.4

Example 27   (a) 1,2-Dimethylbenzene (o-xylene) (C$_8$H$_{10}$)

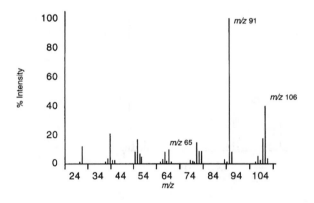

m/z	%	m/z	%	m/z	%
26	1.3	61	1.2	79	8.5
27	11.5	62	2.9	89	2.6
37	1.3	63	7.7	90	1.0
38	3.3	64	1.6	91	100.0
39	20.6	65	9.7	92	7.6
40	2.2	66	1.1	102	1.3
41	2.4	74	2.0	103	5.1
50	7.8	75	1.7	104	2.5
51	16.9	76	1.3	105	17.3
52	6.9	77	14.8	106	39.9
53	4.3	78	8.4	107	3.3

Example 27   (b) 1,3-Dimethylbenzene (*m*-xylene)

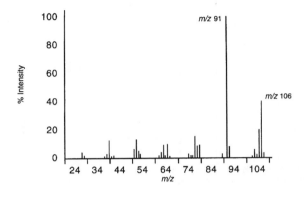

m/z	%	*m/z*	%	*m/z*	%
1 5	1.2	5 3	3.0	7 8	8.6
2 7	3.8	6 1	1.4	7 9	8.7
2 8	1.5	6 2	3.8	8 9	3.1
3 7	1.1	6 3	8.8	9 1	100.0
3 8	2.8	6 4	1.8	9 2	8.0
3 9	12.2	6 5	9.5	102	1.4
4 0	1.4	6 6	1.1	103	5.8
4 1	1.5	7 4	2.6	104	2.5
5 0	6.1	7 5	1.6	105	19.6
5 1	12.6	7 6	1.8	106	39.8
5 2	5.1	7 7	15.1	107	3.6

Example 27   (c) 1,4-Dimethylbenzene (*p*-xylene)

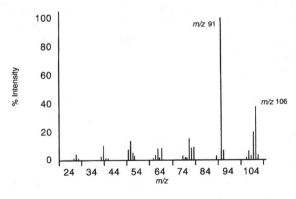

m/z	%	*m/z*	%	*m/z*	%
1 5	1.1	5 3	3.1	7 9	9.1
2 6	1.0	6 1	1.2	8 9	2.9
2 7	3.9	6 2	3.2	9 1	100.0
2 8	1.2	6 3	7.8	9 2	7.0
3 8	2.2	6 4	1.5	102	1.5
3 9	9.9	6 5	8.1	103	6.1
4 0	1.1	7 4	2.8	104	2.8
4 1	1.1	7 5	1.6	105	19.8
5 0	7.2	7 6	1.8	106	37.3
5 1	13.3	7 7	15.0	107	3.3
5 2	4.9	7 8	8.3		

Example 27   (d) Ethylbenzene

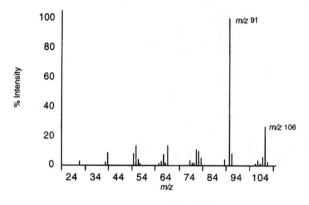

m/z	%	m/z	%	m/z	%
15	1.6	63	7.0	91	100.0
27	2.9	64	1.7	92	8.0
38	2.1	65	13.6	102	1.1
39	8.5	74	3.2	103	3.4
50	7.9	75	1.9	104	1.3
51	13.4	76	1.9	105	5.4
52	4.0	77	10.8	106	26.4
53	1.1	78	9.7	107	2.1
61	1.0	79	5.1		
62	2.8	89	3.8		

In general, the presence of a benzene ring increases the abundance of the molecular ion peaks compared with the saturated analogs. Aromatic ring cleavage usually requires high energies, and significant hydrogen and carbon skeleton rearrangements can occur. Normally, ionization will take place in the π-electron system. This is because it is generally easier to remove a π-electron than it is to remove a σ-electron.

In toluene (Example 26) we see the base peak at m/z 91. This is due to the tropylium ion shown in Figure 4.8.

The tropylium ion is known to be very stable and is formed in preference to the $C_6H_5-CH_2^+$ isomer. The tropylium ion is resonance stabilized, with seven resonance structures. The student may wish to draw these seven structures as an exercise. The tropylium ion is commonly seen in alkylbenzenes (Examples 27a–d), and corresponding ions will also be observed in other benzene derivatives such as the phenols.

For the alkylbenzenes, the ring position of the alkyl groups usually has very little influence on the appearance of the spectra, and isomers such as the xylenes and ethylbenzene are often indistinguishable.

Figure 4.8

5 Halogenated Hydrocarbons

Example 28   (a) Chloromethane (CH₃Cl)

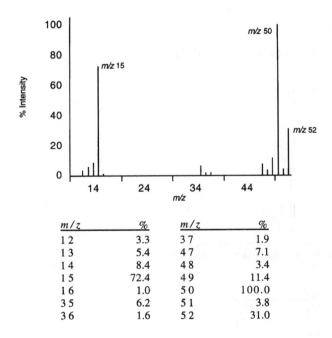

m/z	%	m/z	%
1 2	3.3	3 7	1.9
1 3	5.4	4 7	7.1
1 4	8.4	4 8	3.4
1 5	72.4	4 9	11.4
1 6	1.0	5 0	100.0
3 5	6.2	5 1	3.8
3 6	1.6	5 2	31.0

Example 28   (b) Dichloromethane (methylene chloride) (CH_2Cl_2)

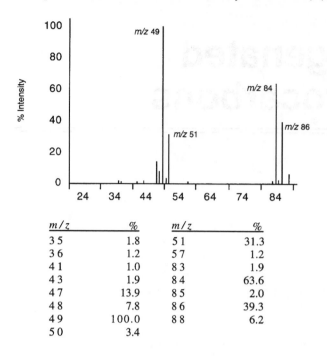

m/z	%	m/z	%
3 5	1.8	5 1	31.3
3 6	1.2	5 7	1.2
4 1	1.0	8 3	1.9
4 3	1.9	8 4	63.6
4 7	13.9	8 5	2.0
4 8	7.8	8 6	39.3
4 9	100.0	8 8	6.2
5 0	3.4		

Example 28   (c) Trichloromethane (chloroform) ($CHCl_3$)

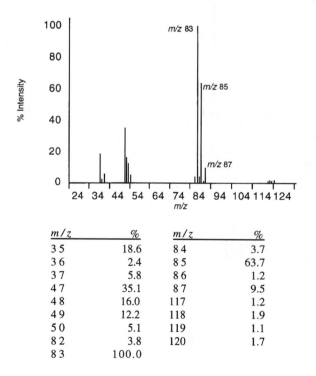

m/z	%	m/z	%
3 5	18.6	8 4	3.7
3 6	2.4	8 5	63.7
3 7	5.8	8 6	1.2
4 7	35.1	8 7	9.5
4 8	16.0	117	1.2
4 9	12.2	118	1.9
5 0	5.1	119	1.1
8 2	3.8	120	1.7
8 3	100.0		

Example 28   (d) Tetrachloromethane (carbon tetrachloride) (CCl$_4$)

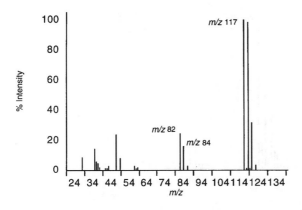

m/z	%	m/z	%	m/z	%
2 8	8.3	4 7	23.4	8 6	2.7
3 5	14.0	4 9	8.0	117	100.0
3 6	5.7	5 7	2.8	118	1.4
3 7	4.4	5 8	1.4	119	98.1
3 8	1.7	5 9	1.7	120	1.1
4 1	1.2	8 2	24.3	121	31.2
4 2	1.1	8 4	15.6	123	3.1
4 3	3.0				

Example 29   (a) 1-Chloropropane (C$_3$H$_7$Cl)

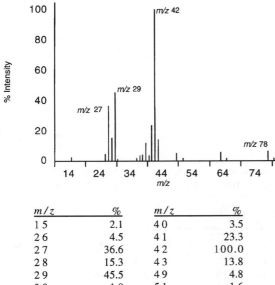

m/z	%	m/z	%
1 5	2.1	4 0	3.5
2 6	4.5	4 1	23.3
2 7	36.6	4 2	100.0
2 8	15.3	4 3	13.8
2 9	45.5	4 9	4.8
3 0	1.0	5 1	1.6
3 6	1.8	6 3	5.5
3 7	3.1	6 5	1.7
3 8	3.8	7 8	6.0
3 9	11.6	8 0	1.9

Example 29   (b) 2-Chloropropane

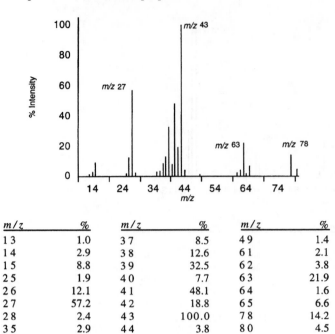

m/z	%	m/z	%	m/z	%
13	1.0	37	8.5	49	1.4
14	2.9	38	12.6	61	2.1
15	8.8	39	32.5	62	3.8
25	1.9	40	7.7	63	21.9
26	12.1	41	48.1	64	1.6
27	57.2	42	18.8	65	6.6
28	2.4	43	100.0	78	14.2
35	2.9	44	3.8	80	4.5
36	3.4				

Example 30   1-Bromobutane (C$_4$H$_9$Br)

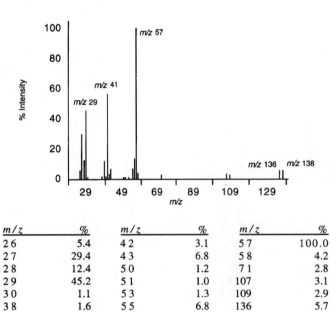

m/z	%	m/z	%	m/z	%
26	5.4	42	3.1	57	100.0
27	29.4	43	6.8	58	4.2
28	12.4	50	1.2	71	2.8
29	45.2	51	1.0	107	3.1
30	1.1	53	1.3	109	2.9
38	1.6	55	6.8	136	5.7
39	11.6	56	13.3	138	5.6
40	1.9				
41	56.4				

Example 31   2,3-Dibromobutane (C$_4$H$_8$Br$_2$)

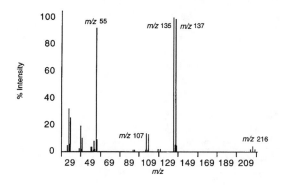

m/z	%	m/z	%	m/z	%
26	4.3	52	1.3	109	13.0
27	31.6	53	7.6	119	1.5
28	5.8	54	1.6	121	1.6
29	25.4	55	92.1	135	100.0
38	2.1	56	9.0	136	4.9
39	18.9	93	1.3	137	98.6
40	1.6	95	1.1	138	4.5
41	9.9	106	1.3	214	1.9
50	3.3	107	13.7	216	3.7
51	3.5	108	1.4	218	1.8

Example 32   (a) 1,2-Dichlorobutane (C$_4$H$_8$Cl$_2$)

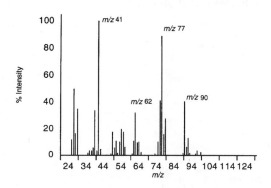

m/z	%	m/z	%	m/z	%
26	12.0	50	5.5	75	9.8
27	50.0	51	10.7	76	40.7
28	16.4	52	1.5	77	89.1
29	34.9	53	10.1	78	15.8
35	1.6	54	19.5	79	27.4
36	3.3	55	17.4	89	1.7
37	3.6	56	6.0	90	39.9
38	5.7	60	1.1	91	6.4
39	33.6	61	10.4	92	12.9
40	3.2	62	32.0	93	1.8
41	100.0	63	9.6	96	1.3
42	4.6	64	10.2	97	3.3
48	1.0	65	2.0	99	2.0
49	17.4	73	1.3		

Example 32   (b) 2,3-Dichlorobutane

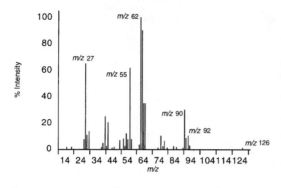

m/z	%	m/z	%	m/z	%
15	1.9	49	6.6	75	10.0
18	1.6	51	8.0	76	2.1
26	7.4	52	2.2	77	6.3
27	65.0	53	11.5	79	1.0
28	10.7	54	7.3	83	2.3
29	13.5	55	61.7	85	1.5
37	1.5	56	7.4	89	1.0
38	4.4	61	3.4	90	29.7
39	24.6	62	100.0	91	8.4
40	2.3	63	90.9	92	10.0
41	20.1	64	34.4	93	2.6
44	1.4	65	34.9	126	1.0
45	1.8	73	1.4		

Example 33   1-Bromo-2-chloro-2-methylpropane (C_4H_8BrCl)

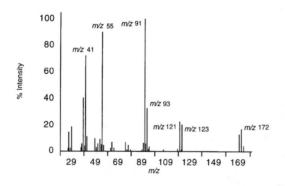

m/z	%	m/z	%	m/z	%
26	2.2	62	2.2	91	100.0
27	14.4	63	6.6	92	5.8
28	2.5	65	2.1	93	32.4
29	18.3	75	6.6	94	1.6
37	2.6	76	1.1	95	3.6
38	5.8	77	4.5	107	1.3
39	40.4	79	1.4	119	1.5
40	4.0	89	1.2	121	22.4
41	72.1	90	6.1	122	1.0
42	11.4	53	9.1	123	20.2
49	9.3	54	4.8	170	12.6
50	2.6	55	89.7	172	16.6
51	2.8	56	4.3	174	4.0

Example 34   3-Chloro-1-hexene (C$_6$H$_{11}$Cl)

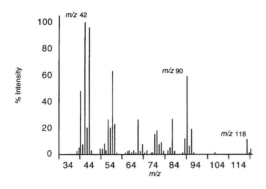

m/z	%	m/z	%	m/z	%
38	2.0	61	1.0	78	9.0
39	5.0	62	2.0	79	2.0
40	48.0	63	3.0	81	3.0
41	7.0	64	1.0	82	5.0
42	100.0	65	3.0	83	27.0
43	20.0	66	1.0	84	2.0
44	96.0	67	26.0	88	1.0
45	3.0	68	2.0	89	12.0
49	4.0	69	7.0	90	59.0
50	4.0	70	1.0	91	6.0
51	8.0	71	3.0	92	19.0
52	3.0	73	1.0	93	1.0
53	26.0	74	1.0	103	1.0
54	20.0	75	15.0	118	11.0
55	63.0	76	18.0	119	1.0
56	23.0	77	7.0	120	4.0
57	1.0				

Example 35   Bromobenzene (C$_6$H$_5$Br)

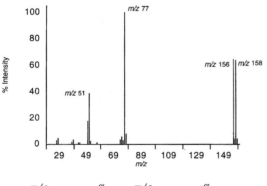

m/z	%	m/z	%
27	2.7	74	3.2
28	4.5	75	5.7
38	1.5	76	5.7
39	3.2	77	100.0
43	1.2	78	7.8
44	1.0	156	64.1
50	17.2	157	4.2
51	38.8	158	63.7
52	2.0	159	3.8
57	1.3		

Example 36   1-Fluorohexane ($C_6H_{11}F$)

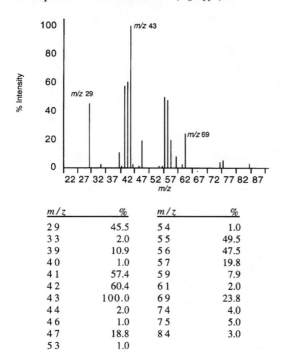

m/z	%	m/z	%
29	45.5	54	1.0
33	2.0	55	49.5
39	10.9	56	47.5
40	1.0	57	19.8
41	57.4	59	7.9
42	60.4	61	2.0
43	100.0	69	23.8
44	2.0	74	4.0
46	1.0	75	5.0
47	18.8	84	3.0
53	1.0		

Example 37   2-Iodobutane (C_4H_9I)

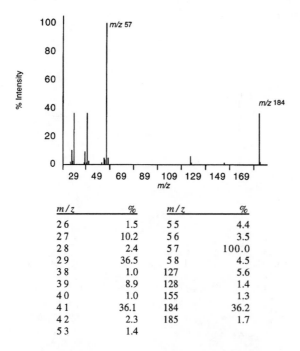

m/z	%	m/z	%
26	1.5	55	4.4
27	10.2	56	3.5
28	2.4	57	100.0
29	36.5	58	4.5
38	1.0	127	5.6
39	8.9	128	1.4
40	1.0	155	1.3
41	36.1	184	36.2
42	2.3	185	1.7
53	1.4		

We can start our study of the halogenated compounds by dividing them into two convenient groups; those that contain F and I (A elements) and those that contain Cl and Br (A + 2 elements).

Compounds containing Cl and Br will exhibit characteristic lines separated by 2 amu, corresponding to ^{35}Cl and ^{37}Cl or ^{79}Br and ^{81}Br. For monohalogenated compounds, the ratio of ^{35}Cl:^{37}Cl will always be 3:1 (3.08:1 exactly) and the ratio of ^{79}Br:^{81}Br will always be 1:1 (1.02:1 exactly). With relatively little practice the student should be able to identify the spectra of monochlorinated or monobrominated compounds by inspection.

It is possible to get an estimate of the number of chlorine or bromine (or both) atoms present in the compound (or ion) by counting the isotope lines in the spectrum and subtracting 1. A monochlorinated compound should exhibit two isotope lines, a dichlorinated compound should exhibit three lines and so on. While this is a generally useful procedure, the student is advised that caution is required when using it. It is often the case that some of the higher mass lines will have very low intensities and may not be visible in the spectrum. This is most commonly encountered with the spectra of polychlorinated (or polybrominated) compounds.

The spectra of polyhalogenated compounds are, of course, much more complicated than those of monohalogenated compounds. One useful tool is a table giving the relative distributions expected from various combinations of bromine and chlorine, and the student may find Table 4 to be helpful. Table 4 is not exhaustive and the student may need a simple method for calculating expected isotopic distributions in polyhalogenated materials. Fortunately, there is a simple way of predicting the pattern exhibited by molecules containing multiple chlorine and bromine atoms, or containing chlorine and bromine in some combination. For predicting these patterns, we need only concern ourselves with the total number of chlorine and bromine atoms.

Consider a molecule containing two chlorine and one bromine atoms. From the table we see that Cl_2 has the pattern 9:6:1, while Br has the pattern 1:1. We can predict the pattern produced by two chlorine atoms and one bromine atom as the product of the

Table 4 Relative distribution of isotope lines for polyhalogenated compounds. This table is representative and not exhaustive

Compound	(M)	(M + 2)	(M + 4)	(M + 6)	(M + 8)	(M + 10)
Cl_2	9	6	1			
ClBr	3	4	1			
Br_2	1	2	1			
Cl_3	27	27	9	1		
Cl_2Br	9	15	7	1		
$ClBr_2$	3	7	5	1		
Br_3	1	3	3	1		
Cl_4	81	108	54	12	1	
Br_4	1	4	6	4	1	
Cl_5	243	405	270	90	15	1
Br_5	1	5	10	10	5	1

individual patterns and represent this as

$$
\begin{aligned}
&(9{:}6{:}1)\ (1{:}1) \\
&(9{:}6{:}1) \times 1 = 9{:}6{:}1 \\
&(9{:}6{:}1) \times 1 = \underline{\quad 9{:}6{:}1} \\
&\qquad\qquad\qquad 9{:}15{:}7{:}1
\end{aligned}
$$

The expected pattern should have four lines, separated by intervals of 2 amu and in a relative ratio of $9{:}15{:}7{:}1$.

Similarly a molecule containing four chlorine atoms can be viewed as $Cl_2 \times Cl_2$

$$
\begin{aligned}
&(9{:}6{:}1) \times (9{:}6{:}1) \\
&(9{:}6{:}1) \times 9 = 81{:}54{:}9 \\
&(9{:}6{:}1) \times 6 = \quad 54{:}36{:}6 \\
&(9{:}6{:}1) \times 1 = \underline{\qquad\qquad 9{:}6{:}1} \\
&\qquad 81{:}108{:}54{:}12{:}1 \text{ for the total pattern.}
\end{aligned}
$$

While in principle this process can be used for any combination of isotopes, most analysts find that it works best for polyhalogenated compounds. If necessary, other $A+2$ elements can be included when appropriate.

When compounds contain many $A+2$ elements, the spectra may not exhibit all of the lines indicated by the calculation above. In the spectrum of a tetrachlorinated hydrocarbon, for example, we may not be able to observe the five lines indicated by the calculation. The higher mass members of the series are often of low intensity and may be missing from the spectrum. We can still determine the total number of chlorines present by comparing the ratios observed in the mass spectrum with the predicted ratios.

For example, we have a spectrum containing three lines separated by 2 amu. How do we know that the compound contains four chlorine atoms (as opposed to two bromine atoms)? From our calculation of the expected distribution for four chlorine atoms, we should observe relative distributions of $81{:}108{:}54{:}12{:}1$. If we re-normalize these distributions, setting the 108 value to 100%, we would have relative intensities of 75%, 100%, 50%, 11% and 0.9% respectively. We can perform a similar calculation for the expected distributions of two bromine atoms and determine relative intensity values of 50%, 100% and 50%.

Finally, we would calculate the relative intensities of the three lines in our spectrum. If the intensities of the lines in our spectrum are reasonably close to the relative intensities predicted for four chlorine atoms, then we can conclude that we have a tetrachlorinated compound, even though two of the lines expected in the spectrum may be missing.

Let us do one more example so that this procedure is clear. Part of the spectrum of a chlorinated hydrocarbon contains the following lines:

m/z	Abundance
290	38.0%
292	75.0%
294	62.0%
296	28.0%
298	7.0%

How many chlorine atoms are actually present?

We know that we have at least four chlorine atoms, since the number of isotope lines will always be one greater than the number of chlorine atoms (the same is true for bromine and generally true for all $A + 2$ elements). If we re-normalize the data, setting the highest abundance to 100%, we get the following values:

m/z	% Abundance	Re-normalized
290	38.0%	50.7%
292	75.0%	100.0%
294	62.0%	82.7%
296	28.0%	37.3%
298	7.0%	9.3%

Is this pattern consistent with what we would expect from a compound with four chlorine atoms? Consulting Table 4, we see that the relative ratios should be:

Line	Relative distribution	Re-normalized
M	81	75.0%
M + 2	108	100.0%
M + 4	54	50.0%
M + 6	12	11.1%
M + 8	1	0.9%

A comparison of the re-normalized values clearly indicates that the compound contains more than four chlorine atoms.

Does our compound contain five chlorine atoms? Using the data from Table 4 and performing similar calculations, we get the following results:

Line	Relative distribution	Re-normalized
M	243	60.0%
M + 2	405	100.0%
M + 4	270	66.7%
M + 6	90	22.2%
M + 8	15	3.7%
M + 10	1	0.2%

And once again the predicted and observed values are significantly different.

What about six chlorine atoms? Consulting Table 4, we find that we do not have tabulated values for six chlorine atoms. Fortunately, the calculation is a simple one and we arrive at the following values:

Line	Relative distribution	Re-normalized
M	729	50.0%
M + 2	1458	100.0%
M + 4	1215	83.3%
M + 6	540	37.0%
M + 8	135	9.3%
M + 10	18	1.2%
M + 12	1	0.06%

A comparison of the re-normalized intensities for m/z 290–298 compares very favorably with the intensities predicted for the 'M'–'M + 8' values. We can conclude that our compound contains six chlorine atoms.

This example was very simple, since we limited our A + 2 elements to chlorine. If one or more bromine atoms may be present, the number of possible combinations increases and so does the amount of work involved in calculating the various possible distributions.

What about the A elements, fluorine and iodine? Of the two, the spectra of compounds containing iodine are generally much easier to identify and interpret than those containing fluorine. This is due to the relative ease with which iodine is lost and the correspondingly large neutral mass loss (127 amu). The spectra of fluorinated compounds are probably the most difficult of the halogenated hydrocarbon spectra to interpret. Part of this difficulty is due to the fact that many fluorinated compounds do not exhibit a molecular ion in their spectra.

The molecular ions of alkyl halides are normally measurable if the halogen is Br or I. When Cl is present, the intensity of the molecular ion is going to depend on several factors including the number of Cl atoms, the length of the carbon chain and the amount of branching in the molecule. Generally, polychlorinated alkanes will not show a significant molecular ion peak. Significant branching will also reduce the intensity of the molecular ion peak, as will increasing chain length. Normally, these factors do not affect the brominated or iodinated compounds to the same extent as the chlorinated analogs.

Consider spectra 28a–d. We can easily identify the molecular ion in chloromethane (28a) and in dichloromethane (28b). In trichloromethane (28c) the molecular ion has a very low intensity and we have to be careful not to overlook it, while in tetrachloromethane (28d) the molecular ion is absent. This series of spectra illustrates the effects of multiple chlorines in a very straightforward fashion, as those effects due to branching and chain length have been eliminated.

In general, the fluorinated hydrocarbons will not show a molecular ion peak (spectrum 36). This makes fluorinated materials among the most difficult to identify, and certainly the determination of the molecular formulae for such compounds is quite difficult or impossible unless other stabilizing functional groups are present.

Halogenated aromatics will show the same trend described above, but in general the intensity of the molecular ion peak is greater for aromatic compounds than for aliphatic compounds (compare spectrum 35, bromobenzene with spectrum 30, 1-bromobutane). While this increase in stability will allow a significant molecular ion peak to be observed, the molecular ion peak for fluorinated aromatics may still be very weak or non-existent.

As a general rule, the influence of halogens on fragmentation reactions is minimal. Iodine has the greatest influence owing to the relative weakness of carbon–iodine bonds. Normally, ionization will occur by removal of an electron from the halogen atom. The general trend observed in ionization is

$$n > \pi > \sigma$$

where n is a non-bonding electron, π is a pi-bonding electron and σ is a sigma-bonding electron.

The most important cleavage in halogenated aliphatic compounds is loss of the halogen as a radical, producing a carbocation. This carbocation will then fragment further as described in Chapter 4. This fragmentation mechanism is most important for iodo and

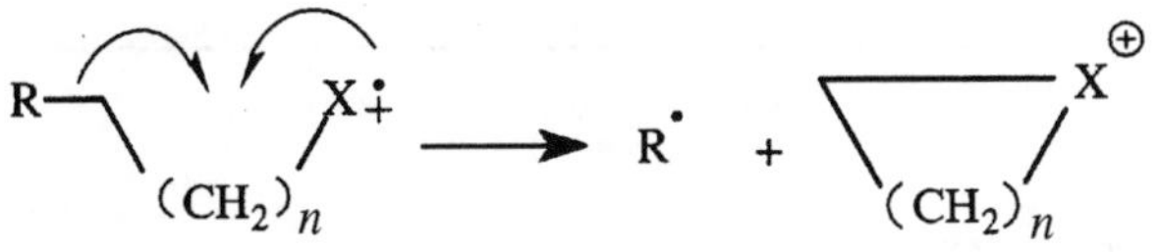

Figure 5.1

bromo compounds, since iodine and bromine are relatively good leaving groups. In chloro and fluoro compounds, loss of the halogen is less likely and the corresponding lines are less intense, although there are exceptions in the case of smaller molecules (Example 29b).

Chloro and fluoro compounds will often lose HX, producing a new odd-electron ion (see m/z 90 in Examples 32a and b). If the compound is a secondary or tertiary halo compound, this loss is much easier and the odd-electron ion produced will have greater intensity.

The α-cleavage mechanism, forming $CH_2=X^+$ and an alkyl radical, is generally less important for the halogenated hydrocarbons. In the spectra of chlorinated and brominated hydrocarbons, loss of an alkyl radical and formation of a five-membered ring is an important exception, provided that the carbon chain is longer than C_4. This mechanism is shown in Figure 5.1.

Compounds containing F and Cl tend to show loss of HF (loss of 20 amu) and HCl (loss of 36 and 38 amu), while loss of HBr and HI is relatively rare. Compounds containing F or I will also typically show loss of 19 or 127 respectively. Frequently, this is the only way of demonstrating the presence of either of these elements.

6 Alcohols

Example 38 Methanol (CH₄O)

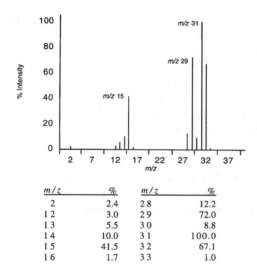

m/z	%	m/z	%
2	2.4	2 8	12.2
1 2	3.0	2 9	72.0
1 3	5.5	3 0	8.8
1 4	10.0	3 1	100.0
1 5	41.5	3 2	67.1
1 6	1.7	3 3	1.0

Example 39 Ethanol (C₂H₆O)

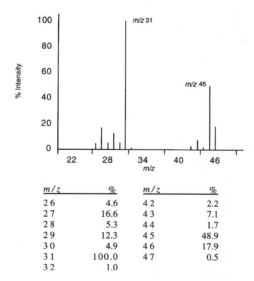

m/z	%	m/z	%
2 6	4.6	4 2	2.2
2 7	16.6	4 3	7.1
2 8	5.3	4 4	1.7
2 9	12.3	4 5	48.9
3 0	4.9	4 6	17.9
3 1	100.0	4 7	0.5
3 2	1.0		

Example 40   2-Butanol ($C_4H_{10}O$)

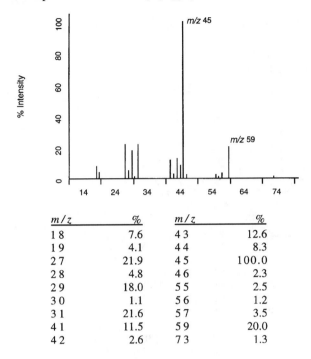

m / z	%	m / z	%
1 8	7.6	4 3	12.6
1 9	4.1	4 4	8.3
2 7	21.9	4 5	100.0
2 8	4.8	4 6	2.3
2 9	18.0	5 5	2.5
3 0	1.1	5 6	1.2
3 1	21.6	5 7	3.5
4 1	11.5	5 9	20.0
4 2	2.6	7 3	1.3

Example 41   2-Methyl-2-propanol ($C_4H_{10}O$)

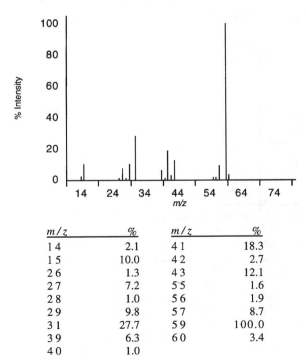

m / z	%	m / z	%
1 4	2.1	4 1	18.3
1 5	10.0	4 2	2.7
2 6	1.3	4 3	12.1
2 7	7.2	5 5	1.6
2 8	1.0	5 6	1.9
2 9	9.8	5 7	8.7
3 1	27.7	5 9	100.0
3 9	6.3	6 0	3.4
4 0	1.0		

Example 42   3-Hexanol ($C_6H_{14}O$)

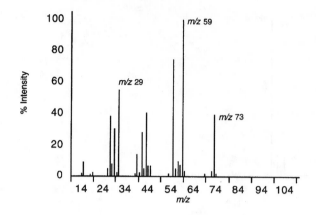

m/z	%	m/z	%	m/z	%
14	1.9	38	1.6	56	5.3
15	8.7	39	14.2	57	9.6
18	1.3	40	2.2	58	7.5
19	2.3	41	28.1	59	100.0
26	4.8	42	5.3	60	3.4
27	38.2	43	40.3	69	1.5
28	8.1	44	6.9	72	3.6
29	30.4	45	6.9	73	38.9
30	2.1	53	1.9	74	1.8
31	54.8	55	74.4		

Example 43   2-Methyl-2-pentanol ($C_6H_{14}O$)

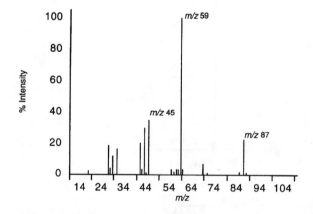

m/z	%	m/z	%	m/z	%
18	2.3	43	29.8	59	100.0
27	18.4	44	1.1	60	3.4
28	4.1	45	34.4	69	6.7
29	11.6	55	3.2	71	1.1
31	16.4	56	1.7	85	1.8
41	20.2	57	3.1	87	22.4
42	3.3	58	3.2	88	1.2

Example 44   Cyclohexanol (C$_6$H$_{12}$O)

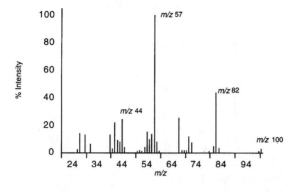

m/z	%	m/z	%	m/z	%
26	2.2	50	1.1	68	1.7
27	14.0	51	1.5	69	1.7
29	13.1	52	1.1	70	1.6
31	6.4	53	3.7	71	11.9
39	13.1	54	15.2	72	7.4
40	2.8	55	9.4	79	1.3
41	21.8	56	13.5	81	4.6
42	8.8	57	100.0	82	43.5
43	8.0	58	7.8	83	3.4
44	24.2	59	1.1	100	2.8
45	3.9	67	25.0		

Example 45   3-Methylcyclohexanol (C$_7$H$_{14}$O)

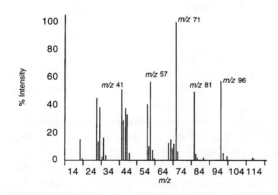

m/z	%	m/z	%	m/z	%
18	14.9	45	5.3	72	6.0
19	1.4	55	40.2	81	49.2
27	44.7	56	9.8	82	4.5
28	13.3	57	56.2	83	1.5
29	38.2	58	7.0	85	1.0
30	2.1	59	1.4	86	1.5
31	16.1	67	12.2	96	57.2
32	3.6	68	15.3	97	5.1
41	50.6	69	8.2	99	2.6
42	28.7	70	11.8	113	1.8
43	37.3	71	100.0	114	1.1
44	32.8				

Example 46   Benzyl alcohol (C$_7$H$_8$O)

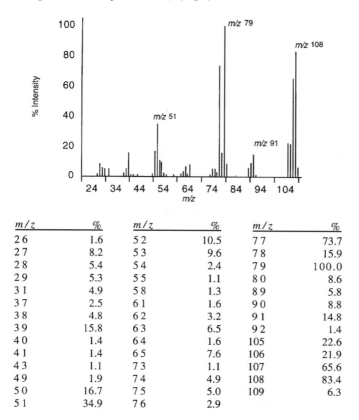

m/z	%	m/z	%	m/z	%
26	1.6	52	10.5	77	73.7
27	8.2	53	9.6	78	15.9
28	5.4	54	2.4	79	100.0
29	5.3	55	1.1	80	8.6
31	4.9	58	1.3	89	5.8
37	2.5	61	1.6	90	8.8
38	4.8	62	3.2	91	14.8
39	15.8	63	6.5	92	1.4
40	1.4	64	1.6	105	22.6
41	1.4	65	7.6	106	21.9
43	1.1	73	1.1	107	65.6
49	1.9	74	4.9	108	83.4
50	16.7	75	5.0	109	6.3
51	34.9	76	2.9		

Alcohols do not generally show an intense molecular ion peak except for very low molecular weight members (methanol and ethanol). This will make the determination of the appropriate molecular formula very difficult or impossible.

Ionization generally occurs by removal of an electron from the oxygen atom, following the trend $n > \pi > \sigma$. The spectra of alcohols are characterized by α-cleavage reactions. Unlike other isomers we have encountered, many isomeric alcohols can be distinguished by their mass spectra. Consider the spectra of 2-butanol (Example 40) and 2-methyl-2-propanol (Example 41). Both of these compounds have a line at m/z 59 which is due to loss of CH$_3$ from the molecular ion. While this line is moderately intense in 2-butanol, it is the base peak in 2-methyl-2-propanol. The base peak in 2-butanol is at m/z 45, which is due to the loss of C$_2$H$_5$ from the molecular ion (Stevenson's rule). These fragmentations are shown in Figures 6.1 and 6.2.

In general, tertiary alcohols will contain the largest total abundance of oxygen-containing ions in their spectra, followed by secondary alcohols, with primary alcohols containing the smallest total abundance of these ions. Normally, the process forming these ions is an initial α-cleavage followed by hydrogen transfer and the elimination of a neutral molecule with the formula C$_n$H$_{2n}$. One typical reaction scheme is shown in Figure 6.3.

2-Butanol

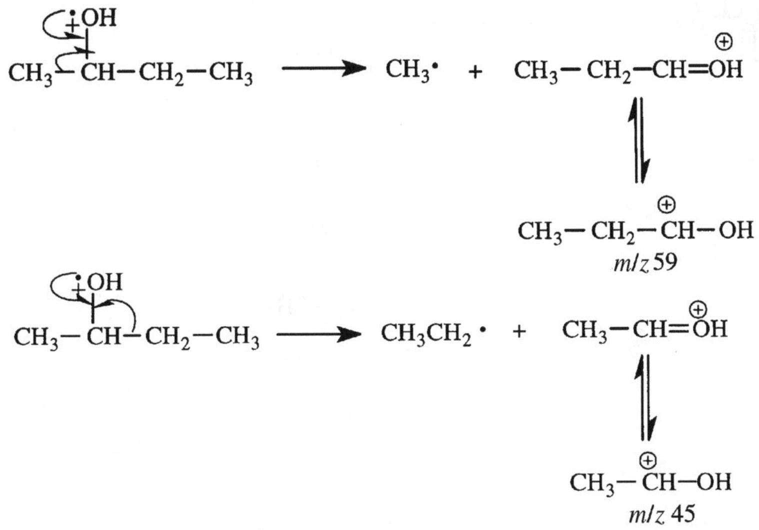

$$CH_3 - \overset{\overset{\displaystyle OH}{|}}{CH} - CH_2 - CH_3 \quad \longrightarrow \quad CH_3\cdot \quad + \quad CH_3 - CH_2 - CH = \overset{\oplus}{O}H$$

$$CH_3 - CH_2 - \overset{\oplus}{CH} - OH$$
$$m/z\,59$$

$$CH_3 - \overset{\overset{\displaystyle OH}{|}}{CH} - CH_2 - CH_3 \quad \longrightarrow \quad CH_3CH_2\cdot \quad + \quad CH_3 - CH = \overset{\oplus}{O}H$$

$$CH_3 - \overset{\oplus}{CH} - OH$$
$$m/z\,45$$

Figure 6.1

2-Methyl-2-propanol

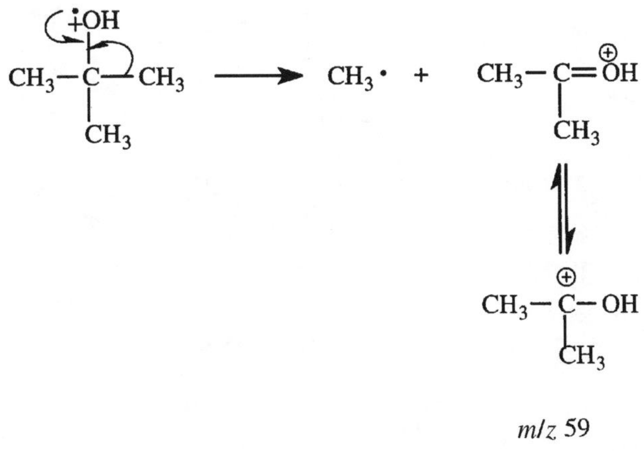

$$CH_3 - \overset{\overset{\displaystyle OH}{|}}{\underset{\underset{\displaystyle CH_3}{|}}{C}} - CH_3 \quad \longrightarrow \quad CH_3\cdot \quad + \quad CH_3 - \overset{\overset{\displaystyle }{|}}{\underset{\underset{\displaystyle CH_3}{|}}{C}} = \overset{\oplus}{O}H$$

$$CH_3 - \overset{\oplus}{\underset{\underset{\displaystyle CH_3}{|}}{C}} - OH$$

$$m/z\,59$$

Figure 6.2

Although the above reaction indicates the initial loss of the largest alkyl radical (Stevenson's rule), similar reactions can occur following the loss of the ethyl or methyl group.

The only other major fragmentation pathways are the loss of H_2O and loss of $(H_2O + C_nH_{2n})$ for $n \geq 2$. These pathways are not particularly important for branched chain alcohols but are significant for the linear alcohols. The mechanism is shown in Figure 6.4. Please note that this mechanism is superficially very similar to a McLafferty

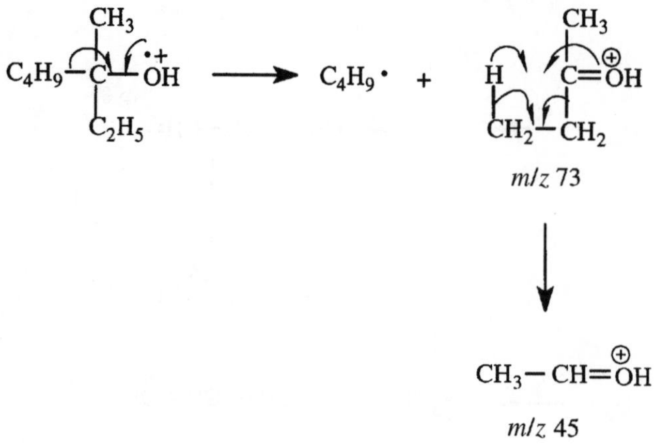

$$C_4H_9-\overset{\overset{\displaystyle CH_3}{|}}{\underset{\underset{\displaystyle C_2H_5}{|}}{C}}-\overset{\bullet +}{O}H \longrightarrow C_4H_9{}^{\bullet} \;+\; \overset{\overset{\displaystyle CH_3}{|}}{\underset{\underset{\displaystyle CH_2-CH_2}{|}}{C}}=\overset{\oplus}{O}H$$

m/z 73

$$CH_3-CH=\overset{\oplus}{O}H$$

m/z 45

Figure 6.3

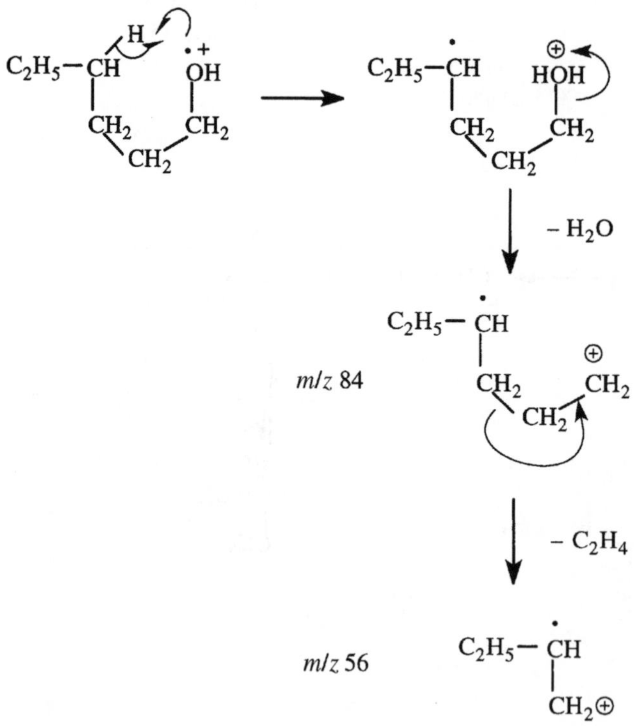

$$C_2H_5-CH \qquad\qquad C_2H_5-\overset{\bullet}{CH} \qquad H\overset{\oplus}{O}H$$

m/z 84

$- H_2O$

$$C_2H_5-\overset{\bullet}{CH}$$

$- C_2H_4$

m/z 56 $C_2H_5-\overset{\bullet}{CH}$, $CH_2\oplus$

Figure 6.4

rearrangement; however, in a true McLafferty rearrangement hydrogen transfer is to an unsaturated heteroatom (for example, a carbonyl oxygen, C=O).

Normally the transfer of hydrogen will take place from the fourth carbon atom, since this forms a six member 'ring' intermediate; transfer can take place from the third carbon but this is less favorable. With alcohols of C_5 or greater, loss of C_2H_4 and H_2O occurs, as a single concerted step and the intermediate ion (*m/z* 84) will show a relatively low intensity.

Notice the masses of the ions formed. These are important OE ions!! The student is reminded that all OE ions are important and whenever such peaks occur in the spectrum the student must do her or his best to identify the fragment ion responsible for them.

Cyclic aliphatic alcohols such as cyclohexanol (Example 44) and 3-methylcyclohex-anol (Example 45) will generally exhibit more intense molecular ions than their acyclic analogs, although the molecular ion lines are still very weak. The first step in the fragmentation process is the opening of the ring structure by α-cleavage (Figure 6.5).

The student should note that this α-cleavage can occur in two different ways. Once the ring is opened, we can have hydrogen transfer or a second α-cleavage eliminating a neutral molecule (Figure 6.6). Both of these reactions are shown below, with α-cleavage producing the fragment on the left and hydrogen transfer producing the fragment on the right.

Note the OE ion at *m/z* 86 formed by elimination of the neutral C_2H_4 molecule. This OE ion can undergo further α-cleavage reactions eliminating C_3H_6 and producing another OE ion at *m/z* 44 ($C_2H_4O^{+\cdot}$). For the ion produced by hydrogen transfer, further α-cleavage produces an EE ion at *m/z* 71 (Figure 6.7).

While we have been following fragmentations resulting from the ring opening between C_1 and C_6, similar reactions can be found for the ring opening between C_1 and C_2.

Loss of water is the other major fragmentation pathway and involves transfer of hydrogen from carbon atoms at the 3, 4 or 5 position in the molecule. This reaction occurs when the molecule is in the boat form transition state and is shown in Figure 6.8 for cyclohexanol. Students should note that in the boat form a six-membered ring transition state is easily achieved.

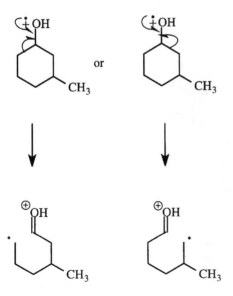

Figure 6.5

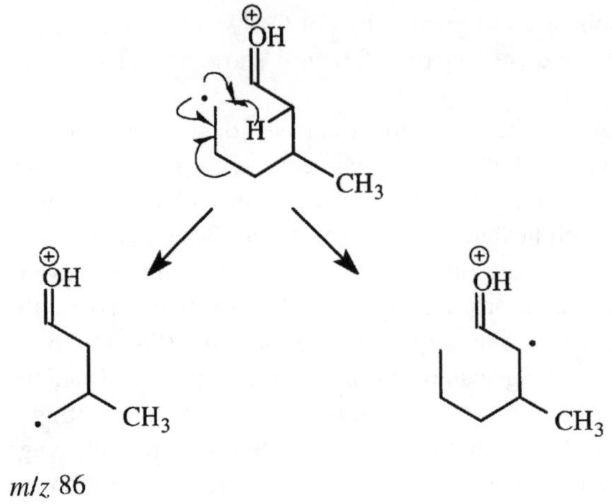

m/z 86

Figure 6.6

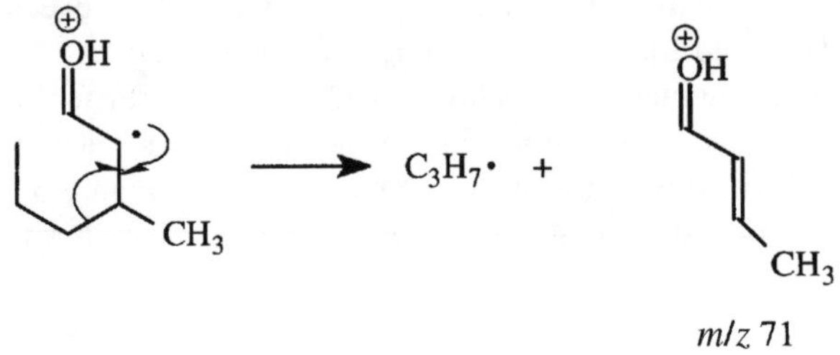

$C_3H_7\cdot$ +

m/z 71

Figure 6.7

m/z 82

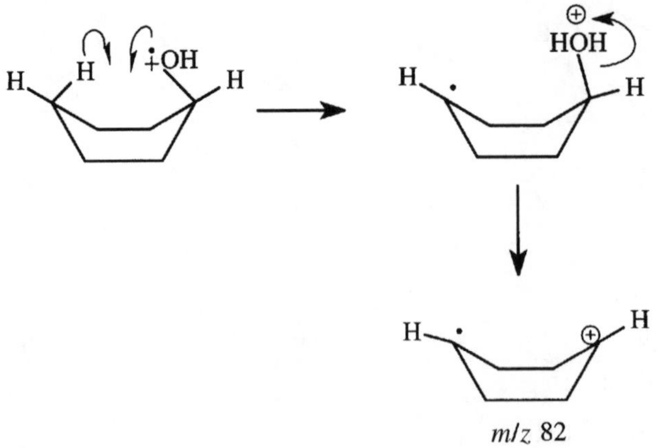

Figure 6.8

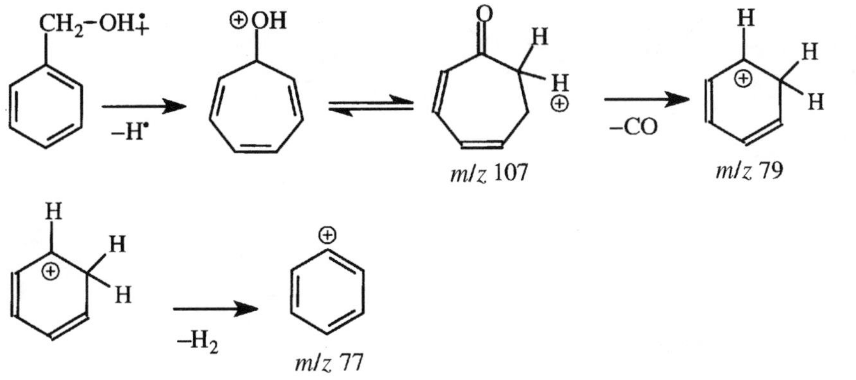

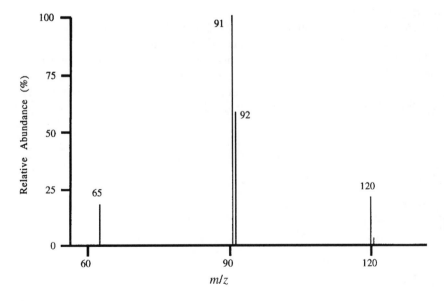

Figure 6.9

As would be expected, the spectra of aromatic alcohols show the most intense molecular ion lines of all alcohols. In benzyl alcohol (Example 46) we see the molecular ion line at m/z 108. Other characteristic lines are at m/z 107 (loss of hydrogen from the parent), m/z 79 (loss of CO from 107) and m/z 77 (loss of H_2 from the m/z 79). These reactions are summarized in Figure 6.9.

The spectrum of 2-phenylethanol is shown below (Figure 6.10). In this spectrum we see a relatively intense molecular ion line at m/z 120. The base peak at m/z 91 is the tropylium ion. This assignment is verified by the line at m/z 65 which corresponds to loss of C_2H_2 from m/z 91. The line at m/z 92 corresponds to loss of CH_2O from the molecular ion. These fragmentation reactions are summarized in Figures 6.11 and 6.12 respectively.

Figure 6.10

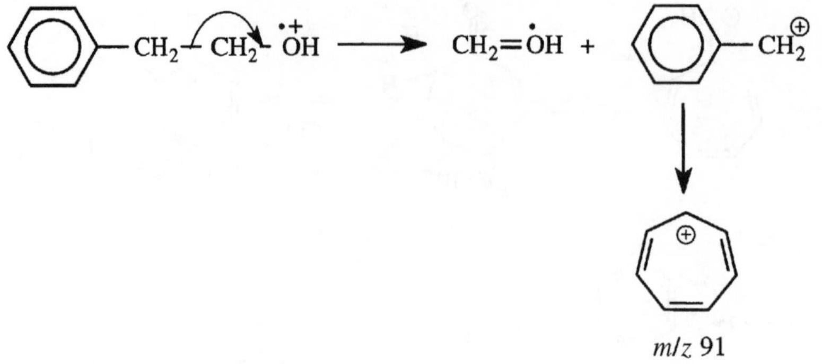

m/z 91

Figure 6.11

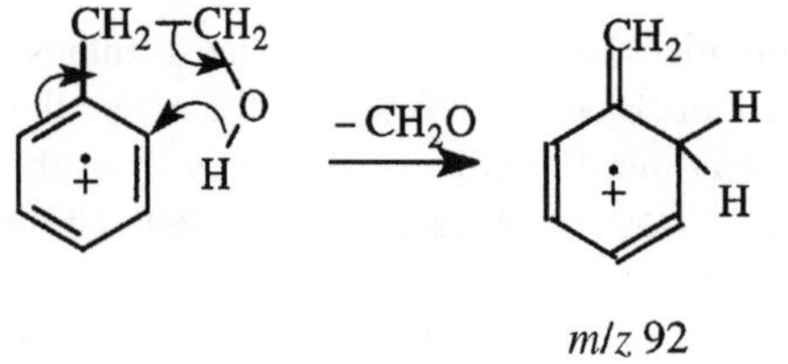

m/z 92

Figure 6.12

A second source of the tropylium ion would be loss of H from *m/z* 92. Note that the hydrogen transfer from the hydroxyl group to the ring is another example of a six-membered ring transition state.

Finally, consider the fragmentation reactions for 2-methylbenzyl alcohol (Figure 6.13). This is an example of the *ortho* effect. The 3- and 4-isomers will also lose water (actually, they will lose OH and H radicals, which amounts to the same thing) but the amounts of the ion at *m/z* 104 produced will be quite different. Once again a six-membered ring transition state is very important. These types of cyclic fragmentation are very important and are not limited to oxygen-containing compounds.

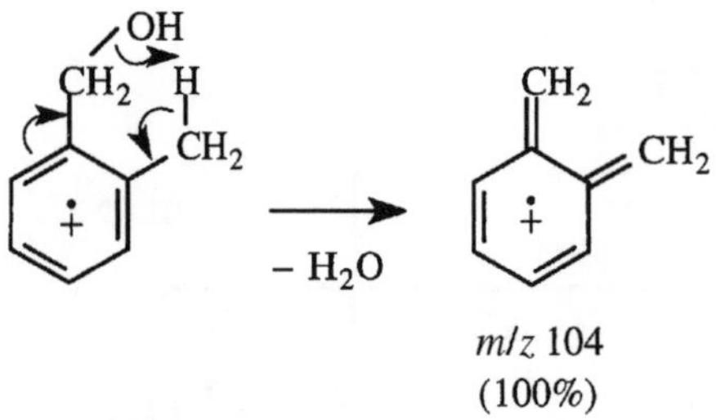

m/z 104
(100%)

Figure 6.13

Alkyl silyl ethers

When analyzing phenols and alcohols, it is often necessary to convert the relatively non-volatile hydroxyl groups with some convenient derivatizing agent. One common method used with these compounds is the formation of trimethylsilyl (TMS) ethers using hexamethyldisilazane:

$$2ROH + [(CH_3)_3Si]_2NH \rightarrow 2ROSi(CH_3)_3 + NH_3$$

These TMS ethers are particularly convenient for both gas chromatography and mass spectroscopy, since the ethers will tend to have higher volatilities than the corresponding alcohols or phenols. When alcohols are derivatized using this reagent, the TMS ether produced generally exhibits a reasonably intense parent ion. This is quite convenient, since the majority of alcohols will not exhibit a usable parent ion, making interpretation quite difficult.

This reagent will also react with free carboxyl and amine groups, producing silyl amines and esters. However, the silyl amines and esters are very sensitive to water and will readily hydrolyze, regenerating the amines and acids. TMS ethers can be formed with primary, secondary or tertiary alcohols and with related compounds such as steroids. With some compounds, significant rearrangements can take place, with the TMS group exhibiting a migration ability similar to that of hydrogen. The spectra of steroids are one case in point.

TMS ethers normally exhibit measurable molecular ions at m/z values 72 higher than the original molecule. TMS ethers will typically produce several characteristic series of fragments. In addition to loss of methyl from the parent molecule ($M - 15$) a reasonably important ion series is shown at m/z 31, 45 and 95. These fragments correspond to ions with the formulae SiH_3^+, $SiCH_5^+$ and $SiC_2H_7^+$ which are produced by elimination of a methyl group and replacement of that group with a hydrogen.

The TMS ethers of phenols are also quite characteristic. In the case of $C_6H_5-O-Si(CH_3)_3$, for example, the parent molecule loses radicals of mass 31 and 33, producing ions corresponding to $C_7H_7OSi^+$ and $C_8H_9Si^+$. These ions are reasonably abundant in the spectrum.

7 Ethers and Phenols

Example 47   Methoxymethane (dimethyl ether) (C_2H_6O)

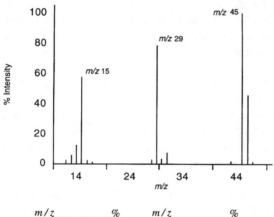

m/z	%	m/z	%
1 2	2.4	2 9	78.9
1 3	5.4	3 0	3.6
1 4	12.5	3 1	7.0
1 5	57.4	4 3	1.7
1 6	2.4	4 5	100.0
1 7	1.1	4 6	45.6
2 8	3.0	4 7	1.2

Example 48   1-Methoxypropane (methyl propyl ether) ($C_4H_{10}O$)

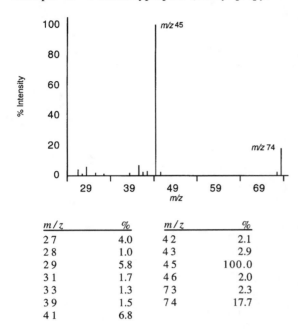

m/z	%	m/z	%
27	4.0	42	2.1
28	1.0	43	2.9
29	5.8	45	100.0
31	1.7	46	2.0
33	1.3	73	2.3
39	1.5	74	17.7
41	6.8		

Example 49   1-Methyl-(1-methylethoxy)ethane (diisopropyl ether) ($C_6H_{14}O$)

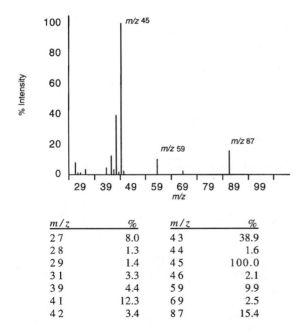

m/z	%	m/z	%
27	8.0	43	38.9
28	1.3	44	1.6
29	1.4	45	100.0
31	3.3	46	2.1
39	4.4	59	9.9
41	12.3	69	2.5
42	3.4	87	15.4

Example 50   1-Propoxypropane (di-n-propyl ether) (C₆H₁₄O)

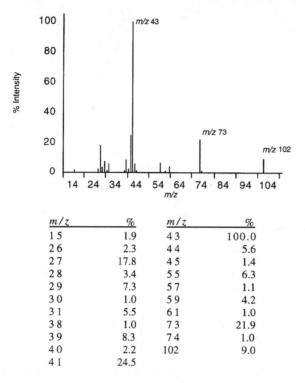

m/z	%	m/z	%
15	1.9	43	100.0
26	2.3	44	5.6
27	17.8	45	1.4
28	3.4	55	6.3
29	7.3	57	1.1
30	1.0	59	4.2
31	5.5	61	1.0
38	1.0	73	21.9
39	8.3	74	1.0
40	2.2	102	9.0
41	24.5		

Example 51   1-Butoxy-2-methylbutane (butyl isopentyl ether) (C₉H₂₀O)

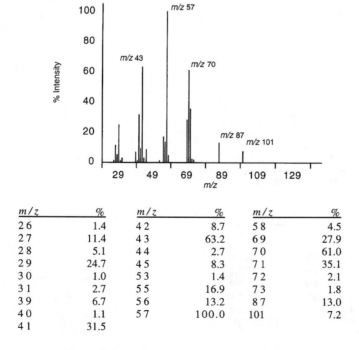

m/z	%	m/z	%	m/z	%
26	1.4	42	8.7	58	4.5
27	11.4	43	63.2	69	27.9
28	5.1	44	2.7	70	61.0
29	24.7	45	8.3	71	35.1
30	1.0	53	1.4	72	2.1
31	2.7	55	16.9	73	1.8
39	6.7	56	13.2	87	13.0
40	1.1	57	100.0	101	7.2
41	31.5				

Example 52   Phenol (C₆H₆O)

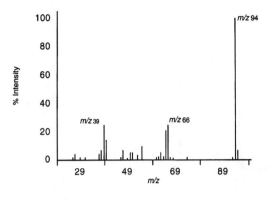

m/z	%	m/z	%	m/z	%
26	1.6	47	6.9	64	2.1
27	4.2	49	1.2	65	20.9
29	1.8	50	5.0	66	24.7
31	1.5	51	5.2	67	1.8
37	4.2	53	3.1	68	1.0
38	6.9	55	9.7	74	1.6
39	24.5	61	1.7	93	1.7
40	14.2	62	2.4	94	100.0
46	1.5	63	5.0	95	6.5

Example 53   3-Methylphenol (C₇H₈O)

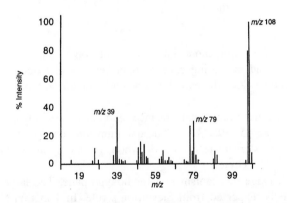

m/z	%	m/z	%	m/z	%
15	2.6	52	8.6	76	1.0
26	2.5	53	14.1	77	26.9
27	11.2	54	5.1	78	8.7
29	2.6	55	3.9	79	30.0
37	6.4	61	3.2	80	6.4
38	12.4	62	4.8	81	2.8
39	32.9	63	9.4	89	3.4
40	3.5	64	2.5	90	9.0
41	2.6	65	2.4	91	6.3
42	1.7	66	4.6	106	2.0
43	2.4	67	2.3	107	80.1
49	2.2	68	1.5	108	100.0
50	12.0	74	2.7	109	7.7
51	15.8	75	1.9		

Example 54   2-Methylphenol (C_7H_8O)

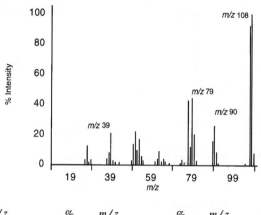

m/z	%	m/z	%	m/z	%
26	3.5	52	9.6	75	1.5
27	12.4	53	16.8	77	42.2
28	1.8	54	5.7	78	11.5
29	3.1	55	2.7	79	44.2
37	3.7	61	2.0	80	20.0
38	7.8	62	4.2	81	2.7
39	20.5	63	9.2	89	15.9
40	2.9	64	2.5	90	25.9
41	1.6	65	3.8	91	8.6
43	1.9	66	2.1	105	1.0
49	2.6	68	1.0	107	91.9
50	13.2	73	1.1	108	100.0
51	21.6	74	3.1	109	7.9

As a class, ethers will tend to exhibit more intense parent ions than do the corresponding alcohols. Branching and increasing carbon chain length can reduce the intensity of the parent ion, and in some cases the parent will be undetected (Examples 49 and 51).

Ethers undergo α-cleavage reactions producing resonance stabilized oxonium ions, resulting in characteristic lines at m/z 45, 59, 73 etc. The oxonium ions produced can undergo a second fragmentation, transferring H from a β-carbon, producing an alkene neutral and $RCH=OH^+$ at m/z 31, 45, 59 etc.

Example 48 illustrates the α-cleavage mechanism in 1-methoxypropane. The largest alkyl (CH_3CH_2) is lost, as would be expected from Stevenson's rule. In 1-methyl-(1-methylethoxy)ethane (Example 49) we do not observe the molecular ion. The fragment at m/z 87 results from α-cleavage with loss of a methyl radical. The base peak at m/z 45 is $CH_3CH=OH^+$ produced by β-H transfer. These reactions are shown in Figure 7.1.

The lines at m/z 43 in 1-methyl-(1-methylethoxy)ethane and 1-propoxypropane (Example 50) are the result of a second fragmentation pathway (Figure 7.2), which eliminates a neutral aldehyde molecule.

The spectrum of 1-propoxypropane is very different from the spectrum of 1-methyl-(1-methylethoxy)ethane (Examples 50 and 49 respectively). In 1-propoxypropane we observe the molecular ion at m/z 102, whereas we cannot see the molecular ion in 1-methyl-(1-methylethoxy)ethane. The intense peak at m/z 73 in 1-propoxypropane is the

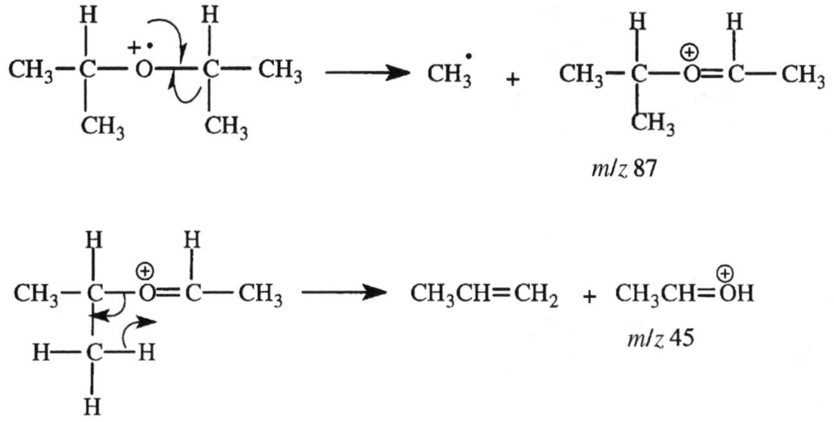

CH₃—C(CH₃)(H)—O⁺·—C(H)(CH₃)—CH₃ ⟶ CH₃• + CH₃—C(H)—O⁺=C(H)—CH₃ (CH₃)

m/z 87

CH₃—C(H)—O⁺=C(H)—CH₃ with H—C(H)H ⟶ CH₃CH=CH₂ + CH₃CH=O⁺H

m/z 45

Figure 7.1

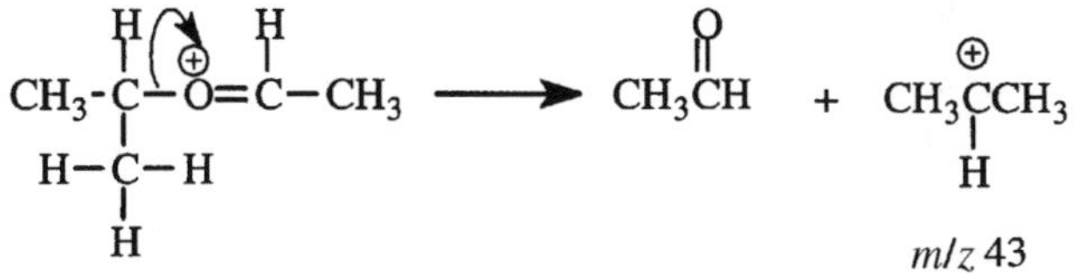

CH₃-C(H)—O⁺=C(H)—CH₃ with H—C(H)—H ⟶ CH₃CHO + CH₃C(O)CH₃ (H)

m/z 43

Figure 7.2

result of α-cleavage with loss of an ethyl radical from either of the n-propyl chains. Finally, the base peak at *m/z* 43 in 1-propoxypropane corresponds to loss of $CH_2=O$ from the fragment ion at *m/z* 73 in a reaction similar to that shown above, producing a propyl cation.

Aromatic ethers behave similarly to aromatic alcohols. Reactions for methoxybenzene (anisole) are typical for this class of compound. The charge tends to stay on the ring as opposed to being on the oxygen molecule. Typically, fragments at *m/z* 77 and 51 are observed as in other simple benzene derivatives. An OE ion at *m/z* 78 can be formed (ionized benzene) by elimination of a neutral aldehyde molecule, in addition to fragments at *m/z* 93, 65 and 39. These reactions are summarized in Figure 7.3.

Phenols (hydroxybenzenes) tend to exhibit very intense molecular ion lines. The spectrum of phenol (Example 52) is dominated by the molecular ion line at *m/z* 94. The line at *m/z* 66 is consistent with loss of CO, indicating that significant rearrangement has occurred. Substituted phenols (Examples 53 and 54) also exhibit very intense molecular ions. The fragment at M − 1 in the spectra of the methylphenols corresponds to loss of hydrogen from the methyl group and rearrangement to form a hydroxy tropylium ion (*m/z* 107). The next fragmentation is loss of CO, producing a line at *m/z* 79. Resonance stabilization of the hydroxy tropylium ion is greatest when the hydroxyl group can participate, and the spectrum of 4-methylphenol (not shown) will have the most intense line at *m/z* 107.

The spectrum of 2-methylphenol exhibits a reasonably intense line at *m/z* 90. This indicates that a molecule of water has been lost through the *ortho*-effect already described. This line is less intense in 3- and 4-methylphenol.

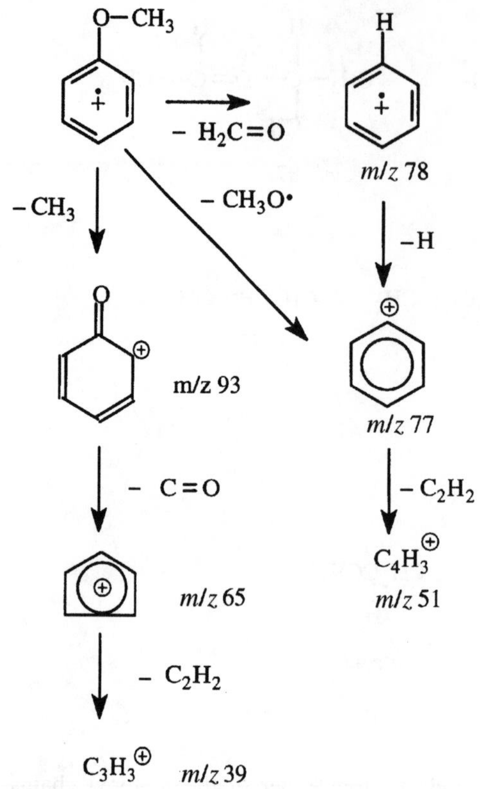

Figure 7.3

Phenols containing longer hydrocarbon chains undergo benzylic cleavage of the chain, resulting in the hydroxy tropylium ion at m/z 107. If the hydrocarbon chain is three carbons long (or longer), transfer of hydrogen from the γ-carbon through a six-membered ring intermediate occurs, producing a line at m/z 108.

8 Aldehydes and Ketones

Example 55   Methanal (formaldehyde) (CH_2O)

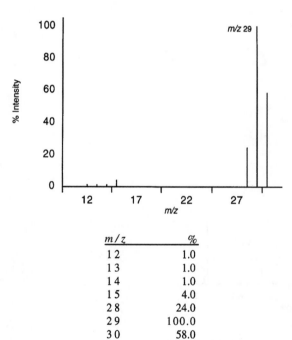

m/z	%
1 2	1.0
1 3	1.0
1 4	1.0
1 5	4.0
2 8	24.0
2 9	100.0
3 0	58.0

Example 56 Ethanal (acetaldehyde) (C_2H_4O)

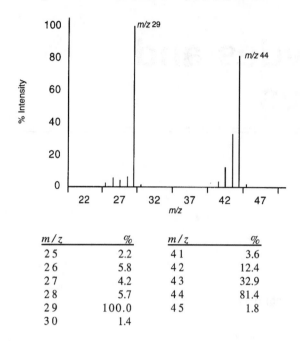

m/z	%	m/z	%
25	2.2	41	3.6
26	5.8	42	12.4
27	4.2	43	32.9
28	5.7	44	81.4
29	100.0	45	1.8
30	1.4		

Example 57 Butanal (C_4H_8O)

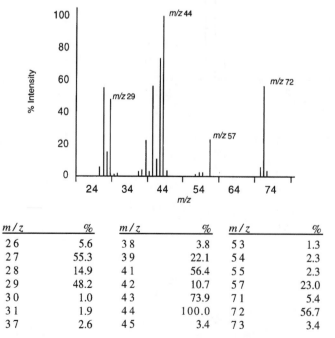

m/z	%	m/z	%	m/z	%
26	5.6	38	3.8	53	1.3
27	55.3	39	22.1	54	2.3
28	14.9	41	56.4	55	2.3
29	48.2	42	10.7	57	23.0
30	1.0	43	73.9	71	5.4
31	1.9	44	100.0	72	56.7
37	2.6	45	3.4	73	3.4

Example 58   Pentanal ($C_5H_{10}O$)

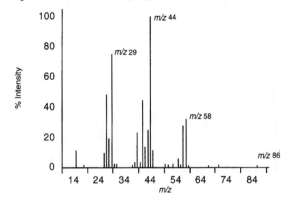

m/z	%	m/z	%	m/z	%
15	11.0	39	22.7	53	2.3
18	1.8	40	3.4	55	5.7
26	9.5	41	44.2	56	1.9
27	48.0	42	13.5	57	27.4
28	19.0	43	24.4	58	31.8
29	74.8	44	100.0	59	1.1
30	2.2	45	11.4	67	1.1
31	2.0	50	2.1	71	1.7
37	1.9	51	1.9	86	1.2
38	3.6				

Example 59   3,3-Dimethylbutanal ($C_6H_{12}O$)

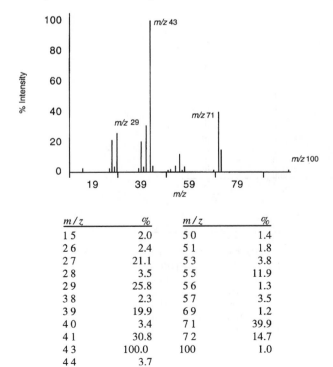

m/z	%	m/z	%
15	2.0	50	1.4
26	2.4	51	1.8
27	21.1	53	3.8
28	3.5	55	11.9
29	25.8	56	1.3
38	2.3	57	3.5
39	19.9	69	1.2
40	3.4	71	39.9
41	30.8	72	14.7
43	100.0	100	1.0
44	3.7		

Example 60   5-Methylhexanal (C$_7$H$_{14}$O)

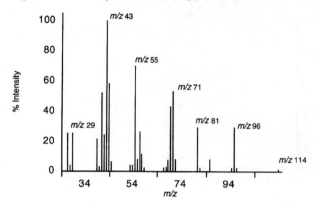

m/z	%	m/z	%	m/z	%
2 7	25.0	5 3	4.0	7 0	43.0
2 8	4.0	5 4	4.0	7 1	53.0
2 9	25.0	5 5	70.0	7 2	8.0
3 9	21.0	5 6	8.0	8 1	29.0
4 0	3.0	5 7	26.0	8 2	2.0
4 1	52.0	5 8	11.0	8 6	8.0
4 2	24.0	5 9	2.0	9 5	2.0
4 3	100.0	6 7	2.0	9 6	29.0
4 4	58.0	6 8	3.0	9 7	2.0
4 5	6.0	6 9	7.0	114	1.0

Example 61   Benzaldehyde (C$_7$H$_6$O)

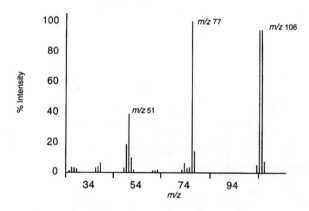

m/z	%	m/z	%	m/z	%
2 6	1.2	5 1	38.4	7 5	3.0
2 7	3.3	5 2	9.3	7 6	3.4
2 8	2.6	5 3	1.7	7 7	100.0
2 9	2.5	6 1	1.3	7 8	13.9
3 7	2.8	6 2	1.4	104	5.0
3 8	3.8	6 3	1.9	105	94.5
3 9	6.3	7 3	1.9	106	94.4
4 9	2.8	7 4	6.2	107	7.2
5 0	18.6				

Example 62   2-Fluorobenzaldehyde (C_7H_5FO)

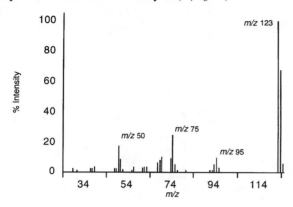

m/z	%	m/z	%	m/z	%
29	2.2	57	3.6	77	1.1
31	1.1	61	2.9	81	1.1
37	2.2	62	3.1	92	1.2
38	2.2	63	3.1	93	1.4
39	3.6	68	6.0	94	5.3
48	2.2	69	7.6	95	9.7
49	2.4	70	9.9	96	2.6
50	17.3	74	9.1	123	100.0
51	8.2	75	24.4	124	67.4
52	1.5	76	4.9	125	5.6
56	1.1				

Example 63   Propanone (acetone) (C_3H_6O)

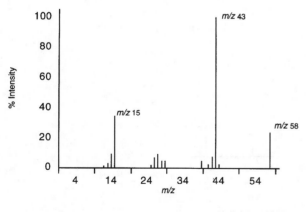

m/z	%	m/z	%
12	1.1	29	4.6
13	2.7	39	4.4
14	8.7	41	2.5
15	34.1	42	7.5
25	1.8	43	100.0
26	6.8	44	2.3
27	8.9	58	23.4
28	4.5		

Example 64   2-Butanone (C₄H₈O)

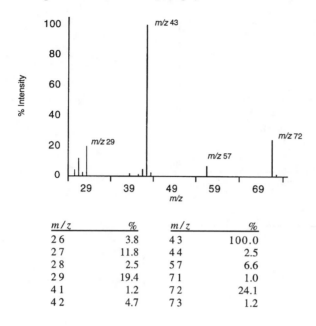

m/z	%	m/z	%
26	3.8	43	100.0
27	11.8	44	2.5
28	2.5	57	6.6
29	19.4	71	1.0
41	1.2	72	24.1
42	4.7	73	1.2

Example 65   2-Pentanone (C₅H₁₀O)

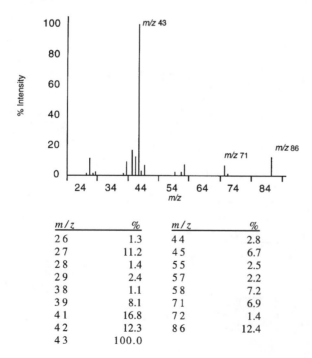

m/z	%	m/z	%
26	1.3	44	2.8
27	11.2	45	6.7
28	1.4	55	2.5
29	2.4	57	2.2
38	1.1	58	7.2
39	8.1	71	6.9
41	16.8	72	1.4
42	12.3	86	12.4
43	100.0		

Example 66   4-Methyl-2-pentanone (methyl isobutyl ketone) ($C_6H_{12}O$)

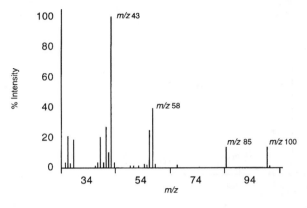

m/z	%	m/z	%	m/z	%
26	3.3	41	26.7	56	1.5
27	20.5	42	9.9	57	24.5
28	3.0	43	100.0	58	39.1
29	18.6	44	3.3	59	2.5
37	1.4	50	1.1	67	1.6
38	3.2	51	1.2	85	13.2
39	20.2	53	1.3	100	13.5
40	3.2	55	2.3	101	1.0

Example 67   4-Heptanone ($C_7H_{14}O$)

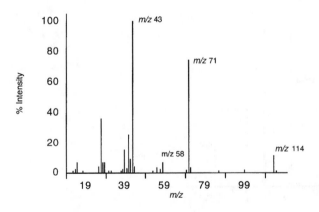

m/z	%	m/z	%	m/z	%
13	1.3	37	1.1	57	2.0
14	2.0	38	2.0	58	6.5
15	6.8	39	15.2	70	1.5
18	1.0	40	2.9	71	74.2
26	3.7	41	25.0	72	3.6
27	35.5	42	9.1	86	1.2
28	6.9	43	100.0	99	1.9
29	6.8	44	3.8	114	11.3
31	1.1	53	1.2	115	1.3
32	1.0	55	3.4		

86

Example 68   2-Octanone ($C_8H_{16}O$)

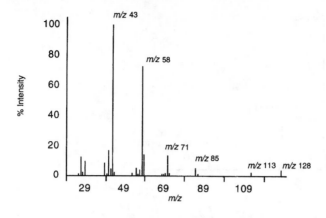

m/z	%	m/z	%	m/z	%
26	1.2	44	2.4	69	1.0
27	12.4	53	1.6	70	1.8
28	2.4	55	5.0	71	13.4
29	9.3	56	1.4	72	1.5
39	8.6	57	4.1	85	5.1
40	1.4	58	72.9	86	1.3
41	16.6	59	13.9	113	2.3
42	4.6	68	1.4	128	3.9
43	100.0				

Example 69   6-Methyl-3-heptanone ($C_8H_{16}O$)

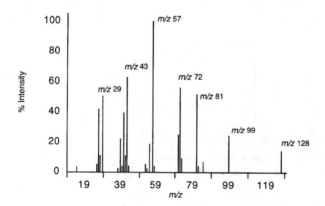

m/z	%	m/z	%	m/z	%
15	3.4	42	10.6	71	24.7
26	5.0	43	92.4	72	55.6
27	41.6	44	4.1	73	8.7
28	10.5	53	5.1	81	51.4
29	50.4	54	2.2	82	3.9
38	2.2	55	18.7	85	6.9
39	21.9	57	100.0	99	24.1
40	4.2	58	4.1	128	13.8
41	39.2				

Example 70   1-Phenyl-2-propanone (phenylacetone) ($C_9H_{10}O$)

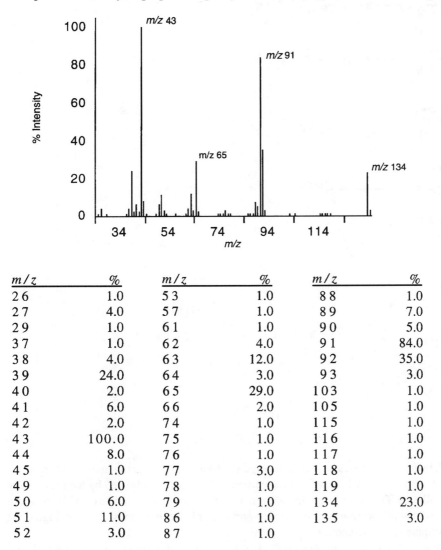

m/z	%	m/z	%	m/z	%
26	1.0	53	1.0	88	1.0
27	4.0	57	1.0	89	7.0
29	1.0	61	1.0	90	5.0
37	1.0	62	4.0	91	84.0
38	4.0	63	12.0	92	35.0
39	24.0	64	3.0	93	3.0
40	2.0	65	29.0	103	1.0
41	6.0	66	2.0	105	1.0
42	2.0	74	1.0	115	1.0
43	100.0	75	1.0	116	1.0
44	8.0	76	1.0	117	1.0
45	1.0	77	3.0	118	1.0
49	1.0	78	1.0	119	1.0
50	6.0	79	1.0	134	23.0
51	11.0	86	1.0	135	3.0
52	3.0	87	1.0		

Aldehydes and ketones usually exhibit a molecular ion, even when some chain branching occurs or the molecule is of considerable size. The α-cleavage mechanism is generally more important for ketones than for aldehydes. In aldehydes, loss of hydrogen from the carbonyl carbon will be important when the acylium ion (RCO^+) can be stabilized (by an aromatic ring, for example). In ketones, α-cleavage can produce either of two acylium ions or two alkyl ions. The more abundant acylium ion is formed by loss of the larger alkyl radical. The more abundant alkyl ion is usually the more stable of the pair of carbocations. These possibilities are illustrated in Figure 8.1.

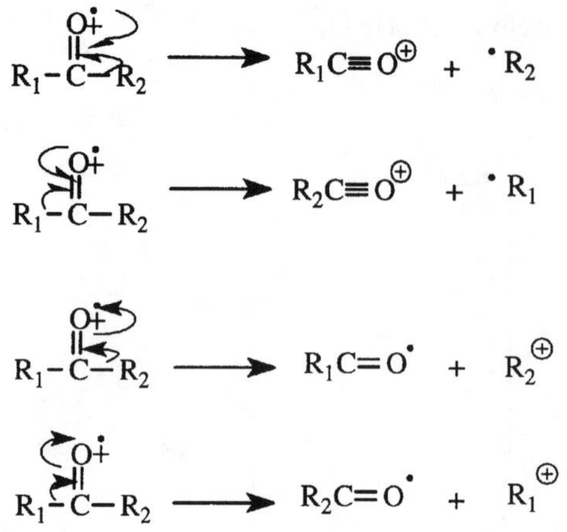

$$R_1-\overset{\overset{\displaystyle O^{+\cdot}}{\|}}{C}-R_2 \longrightarrow R_1C\equiv O^{\oplus} + {}^{\cdot}R_2$$

$$R_1-\overset{\overset{\displaystyle O^{+\cdot}}{\|}}{C}-R_2 \longrightarrow R_2C\equiv O^{\oplus} + {}^{\cdot}R_1$$

$$R_1-\overset{\overset{\displaystyle O^{+\cdot}}{\|}}{C}-R_2 \longrightarrow R_1C=O^{\cdot} + R_2^{\oplus}$$

$$R_1-\overset{\overset{\displaystyle O^{+\cdot}}{\|}}{C}-R_2 \longrightarrow R_2C=O^{\cdot} + R_1^{\oplus}$$

Figure 8.1

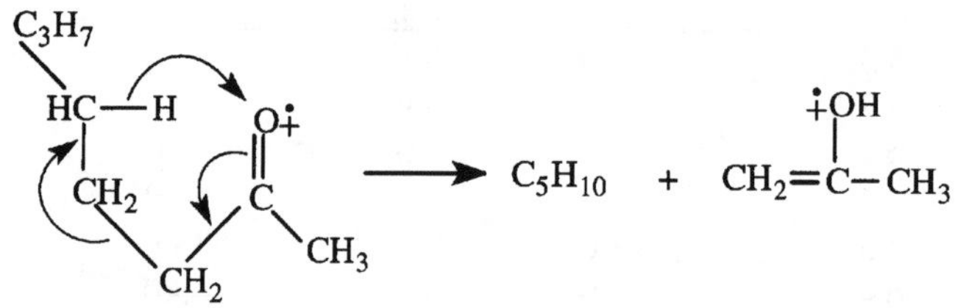

Figure 8.2

Aldehydes and ketones will also undergo the McLafferty rearrangement. This reaction involves transfer of hydrogen from a γ-carbon to the oxygen, followed by the elimination of an alkene. This results in an OE ion being formed, and students are once again reminded that all OE fragment ions are important. This mechanism is shown in Figure 8.2 for Example 68 (2-octanone).

This is such an important reaction that it has been studied in considerable detail, and the following observations have been made:

1. Transfer of hydrogen is always from the γ-carbon. If this carbon is part of a double bond, then the hydrogen will not be transferred.
2. Other groups such as $-S=O$, $-P=O$, $C=C-C=N$ will show the rearrangement. It is not limited to carbonyl compounds.
3. The α, β and γ atoms do not need to be carbon. They can be various combinations of C, N, O and S so long as there is a γ-H that can rearrange. One typical class of compounds meeting this criteria is ethyl esters.

The following list of general compound types and their McLafferty rearrangement peaks may be useful for the student.

Compound type	m/z	Structure
Aldehydes	44	$CH_2{=}\overset{+\!\!\cdot OH}{C}\!-\!H$
Methyl ketones	58	$CH_2{=}\overset{+\!\!\cdot OH}{C}\!-\!CH_3$
Amides	59	$CH_2{=}\overset{+\!\!\cdot OH}{C}\!-\!NH_2$
Acids	60	$CH_2{=}\overset{+\!\!\cdot OH}{C}\!-\!OH$
Ethyl ketones	72	$CH_2{=}\overset{+\!\!\cdot OH}{C}\!-\!CH_2CH_3$
Methyl esters	74	$CH_2{=}\overset{+\!\!\cdot OH}{C}\!-\!OCH_3$
Propyl ketones	86	$CH_2{=}\overset{+\!\!\cdot OH}{C}\!-\!CH_2CH_2CH_3$
	58	$CH_2{=}\overset{+\!\!\cdot OH_2}{C}\!-\!CH_2$
Ethyl esters	88	$CH_2{=}\overset{+\!\!\cdot OH}{C}\!-\!OCH_2CH_3$
	$(M - C_2H_2)$	$CH_2{-}\overset{+\!\!\cdot OH}{C}{=}O$
Phenyl ketones	120	$CH_2{=}\overset{+\!\!\cdot OH}{C}\!-\!C_6H_5$
Phenyl esters	136	$CH_2{=}\overset{+\!\!\cdot OH}{C}\!-\!OC_6H_5$

The student should note that, with the propyl ketones, the McLafferty rearrangement can occur twice consecutively. With the ethyl esters, the rearrangement can occur on either side of the carbonyl with loss of the alkene from either the acid chain or the alcohol chain. The student should also note that the $CH_2=$ groups in all of these ions can be substituted either singly or doubly. This will increase the m/z value of the ion.

The spectrum of ethanal (Example 56) exhibits a strong molecular ion at m/z 44 and a base peak at m/z 29. The base peak corresponds to CHO^+. In butanal and pentanal (Examples 57 and 58) the base peak at m/z 44 is the OE ion formed by the McLafferty rearrangement (CH_2CHOH). In the spectrum of 3,3-dimethylbutanal (Example 59) there is a very small peak at m/z 44, indicating that the McLafferty rearrangement can occur in this molecule. The relatively intense peak at m/z 71 corresponds to loss of CHO from the parent. The base peak at m/z 43 is consistent with fragmentation between the second and third carbons, producing CH_3CO^+ or $C_4H_9^+$. The spectrum of 5-methylhexanal (Example 60) is typical of the larger aldehydes and exhibits the hydrocarbon series of lines at m/z 29, 43, 57 and 71. The lines at m/z 96 and 81 indicate the sequential loss of H_2O and CH_3.

Aromatic aldehydes will have more intense molecular ions than the corresponding aliphatic aldehydes. The spectrum of benzaldehyde (Example 61) has a very intense molecular ion at m/z 106. Other intense lines in the spectrum are at m/z 105 (loss of H from the carbonyl carbon), m/z 77 ($C_6H_5^+$) and m/z 51 ($C_4H_3^+$). The student should recognize the fragments at m/z 51 and 77 as being characteristic of monosubstituted benzenes.

The spectrum of 2-fluorobenzaldehyde (Example 62) is included to demonstrate that not all fluorinated compounds have weak or undetectable parent ions. This will often be the case when there are other more easily ionized functional groups present in fluorinated compounds. The peak at m/z 95 corresponds to loss of CHO from the parent, while m/z 75 is consistent with loss of HF from m/z 95.

Generally speaking, the spectra of ketones will have more intense molecular ions than the corresponding aldehydes, especially as the size of the molecule increases. This can be seen in the comparison of 2-pentanone with pentanal (Examples 65 and 58 respectively), methyl isobutyl ketone with 3,3-dimethylbutanal (Examples 66 and 59 respectively) and 4-heptanone with 5-methylhexanal (Examples 67 and 60 respectively).

The spectrum of 2-butanone (Example 64) exhibits a base peak at m/z 43 which corresponds to CH_3CO^+. The spectrum of 2-pentanone exhibits a weak McLafferty peak at m/z 44. In 4-methyl-2-pentanone (Example 66) we observe a McLafferty peak at m/z 58 and α-cleavage peaks at m/z 85 and 43.

The spectrum of 4-heptanone exhibits relatively weak OE lines at m/z 72 and 58. The expected McLafferty peak at m/z 86 has an intensity of about 1.2%. The peak at m/z 58 is very important because it indicates a double McLafferty rearrangement. The intense line at m/z 71 indicates loss of C_3H_7 from the parent through α-cleavage, while m/z 43 is consistent with loss of CO from m/z 71. The spectrum of 2-octanone (Example 68) exhibits the expected McLafferty peak at m/z 58.

The spectrum of 6-methyl-3-heptanone (Example 69) exhibits the McLafferty peak at m/z 72. Other lines in the spectrum are consistent with hydrocarbon fragmentation and the characteristic lines at m/z 29, 43, 57, 71, 85 and 99 are readily observed. Finally, the spectrum of 1-phenyl-2-propanone (Example 70) exhibits an intense line at m/z 91. In conjunction with the line at m/z 65 we would identify this as a tropylium ion. The base peak at m/z 43 is consistent with CH_3CO^+.

9 Esters and Acids

Example 71   Methyl ethanoate (acetic acid, methyl ester) ($C_3H_6O_2$)

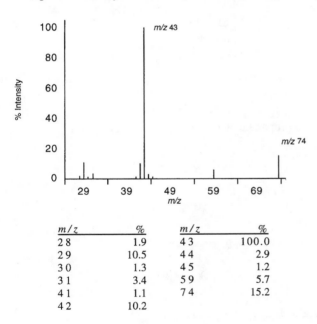

m/z	%		m/z	%
28	1.9		43	100.0
29	10.5		44	2.9
30	1.3		45	1.2
31	3.4		59	5.7
41	1.1		74	15.2
42	10.2			

Example 72   Ethyl methanoate (formic acid, ethyl ester) ($C_3H_6O_2$)

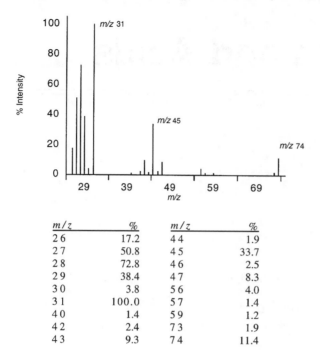

m/z	%	m/z	%
2 6	17.2	4 4	1.9
2 7	50.8	4 5	33.7
2 8	72.8	4 6	2.5
2 9	38.4	4 7	8.3
3 0	3.8	5 6	4.0
3 1	100.0	5 7	1.4
4 0	1.4	5 9	1.2
4 2	2.4	7 3	1.9
4 3	9.3	7 4	11.4

Example 73   Methyl propanoate (propanoic acid, methyl ester) ($C_4H_8O_2$)

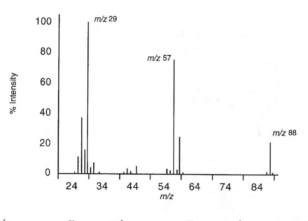

m/z	%	m/z	%	m/z	%
2 5	1.4	3 3	1.1	5 7	75.3
2 6	11.4	4 1	1.1	5 8	2.7
2 7	36.7	4 2	3.1	5 9	24.6
2 8	15.8	4 3	1.9	6 0	1.0
2 9	100.0	4 4	5.0	8 7	1.6
3 0	4.0	5 5	3.4	8 8	21.3
3 1	7.2	5 6	2.4	8 9	1.0

Example 74   Methyl butanoate (butanoic acid, methyl ester) ($C_5H_{10}O_2$)

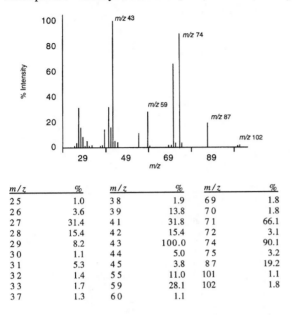

m/z	%	m/z	%	m/z	%
25	1.0	38	1.9	69	1.8
26	3.6	39	13.8	70	1.8
27	31.4	41	31.8	71	66.1
28	15.4	42	15.4	72	3.1
29	8.2	43	100.0	74	90.1
30	1.1	44	5.0	75	3.2
31	5.3	45	3.8	87	19.2
32	1.4	55	11.0	101	1.1
33	1.7	59	28.1	102	1.8
37	1.3	60	1.1		

Example 75   Methyl 2,4,6-trimethyloctanoate (2,4,6-trimethyl octanoic acid, methyl ester) ($C_{12}H_{24}O_2$)

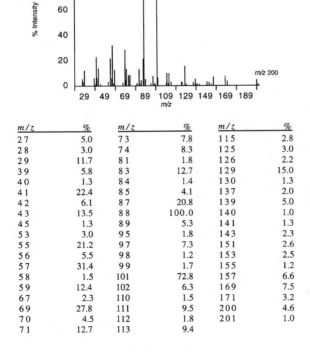

m/z	%	m/z	%	m/z	%
27	5.0	73	7.8	115	2.8
28	3.0	74	8.3	125	3.0
29	11.7	81	1.8	126	2.2
39	5.8	83	12.7	129	15.0
40	1.3	84	1.4	130	1.3
41	22.4	85	4.1	137	2.0
42	6.1	87	20.8	139	5.0
43	13.5	88	100.0	140	1.0
45	1.3	89	5.3	141	1.3
53	3.0	95	1.8	143	2.3
55	21.2	97	7.3	151	2.6
56	5.5	98	1.2	153	2.5
57	31.4	99	1.7	155	1.2
58	1.5	101	72.8	157	6.6
59	12.4	102	6.3	169	7.5
67	2.3	110	1.5	171	3.2
69	27.8	111	9.5	200	4.6
70	4.5	112	1.8	201	1.0
71	12.7	113	9.4		

Example 76   Ethyl decanoate (decanoic acid, ethyl ester) ($C_{12}H_{24}O_2$)

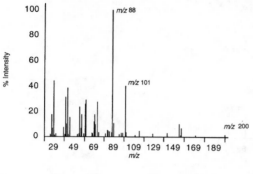

m/z	%	m/z	%	m/z	%
26	1.5	57	17.1	88	100.0
27	17.3	59	2.1	89	10.7
28	6.8	60	25.8	95	1.6
29	44.1	61	28.8	97	2.7
30	1.4	67	2.6	98	2.8
31	2.1	68	2.0	101	40.5
39	7.2	69	12.0	102	3.6
40	1.6	70	17.2	111	1.0
41	31.4	71	9.7	115	4.5
42	10.1	73	27.5	129	2.0
43	38.3	74	3.4	143	2.8
44	1.5	81	2.0	155	10.0
45	15.0	83	5.0	156	1.3
53	1.8	84	4.5	157	6.9
54	2.1	85	4.2	171	1.0
55	23.5	87	3.5	200	1.5
56	6.7				

Example 77   Methyl undecanoate (undecanoic acid, methyl ester) ($C_{12}H_{24}O_2$)

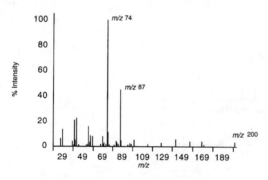

m/z	%	m/z	%	m/z	%
27	6.4	67	1.6	88	4.5
29	13.5	69	7.7	95	1.3
39	4.2	70	1.6	97	2.2
41	20.5	71	2.6	98	1.6
42	4.8	73	1.9	101	4.8
43	22.1	74	100.0	115	1.9
45	1.3	75	11.2	129	2.6
53	1.3	81	1.3	143	5.4
54	1.6	83	3.8	157	4.2
55	15.4	84	2.6	169	4.2
56	3.2	85	1.9	171	1.3
57	8.6	87	44.9	200	3.2
59	8.0				

Example 78   Methyl 3-methylbenzoate (3-methylbenzoic acid, methyl ester) (C$_9$H$_{10}$O$_2$)

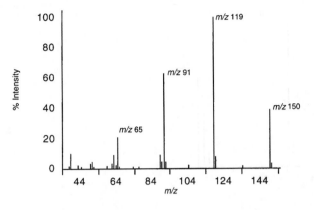

m/z	%	m/z	%	m/z	%
38	1.6	63	9.0	92	4.6
39	10.0	64	2.5	105	2.5
41	2.5	65	20.7	119	100.0
45	1.3	66	1.2	120	7.9
50	3.1	74	1.4	135	1.9
51	4.5	77	1.4	149	1.8
52	1.2	89	9.0	150	38.5
59	1.6	90	4.7	151	3.4
62	3.4	91	62.6		

Example 79   Methyl phenylethanoate (benzeneacetic acid, methyl ester) (C$_9$H$_{10}$O$_2$)

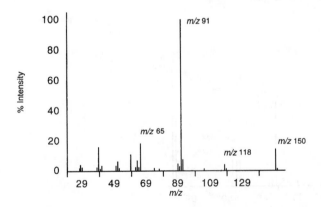

m/z	%	m/z	%	m/z	%
27	1.6	52	1.7	90	2.9
28	3.8	59	10.7	91	100.0
29	2.2	62	2.3	92	7.4
38	2.4	63	6.7	105	1.0
39	15.6	64	2.0	118	4.0
40	1.2	65	17.9	119	1.3
41	3.2	74	1.8	150	14.1
50	3.1	77	1.3	151	1.3
51	5.9	89	4.3		

Example 80   Methanoic acid (formic acid) (CH$_2$O$_2$)

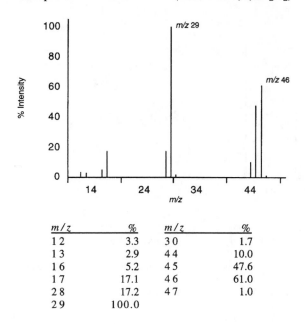

m/z	%	m/z	%
12	3.3	30	1.7
13	2.9	44	10.0
16	5.2	45	47.6
17	17.1	46	61.0
28	17.2	47	1.0
29	100.0		

Example 81   Propanoic acid (C$_3$H$_6$O$_2$)

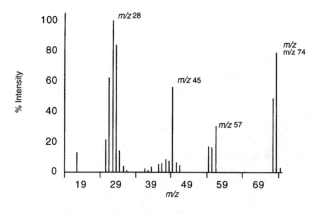

m/z	%	m/z	%	m/z	%
18	13.1	37	2.1	46	6.0
26	21.1	38	1.0	47	4.5
27	61.8	39	3.6	55	16.8
28	100.0	41	5.1	56	16.4
29	83.7	42	5.5	57	30.1
30	14.1	43	8.4	73	48.4
31	4.2	44	7.1	74	78.6
32	1.4	45	55.8	75	2.6

Example 82   Butanoic acid ($C_4H_8O_2$)

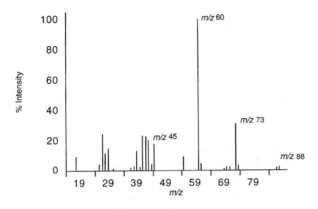

m/z	%	m/z	%	m/z	%
1 8	8.7	4 0	2.2	6 1	4.6
2 6	4.0	4 1	22.8	6 9	1.4
2 7	24.2	4 2	22.1	7 0	2.3
2 8	11.0	4 3	19.9	7 1	2.5
2 9	14.4	4 4	4.1	7 3	31.0
3 1	1.3	4 5	17.1	7 4	3.5
3 7	1.6	5 5	8.9	8 7	1.5
3 8	2.8	6 0	100.0	8 8	2.5
3 9	12.9				

Example 83   Hexanoic acid ($C_6H_{12}O_2$)

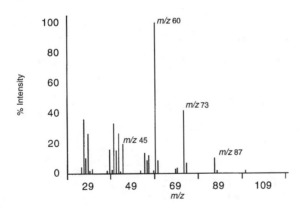

m/z	%	m/z	%	m/z	%
2 6	3.7	4 2	15.2	6 0	100.0
2 7	36.0	4 3	26.5	6 1	8.5
2 8	9.9	4 4	1.2	6 9	2.9
2 9	26.0	4 5	19.8	7 0	3.1
3 0	1.6	5 3	1.6	7 3	41.6
3 1	3.0	5 5	13.6	7 4	6.9
3 8	1.5	5 6	8.2	8 7	10.0
3 9	15.6	5 7	11.9	8 8	1.5
4 0	2.5	5 9	1.6	101	1.8
4 1	32.8				

98   ESTERS AND ACIDS

Example 84   2-Methylpentanoic acid ($C_6H_{12}O_2$)

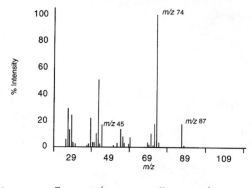

m/z	%	m/z	%	m/z	%
26	5.7	42	9.9	60	7.1
27	29.3	43	51.0	69	3.3
28	12.9	44	2.2	70	1.9
29	23.8	45	16.9	71	10.3
30	3.2	51	1.3	72	1.1
31	2.3	53	2.6	73	17.3
37	1.0	55	13.5	74	100.0
38	2.4	56	8.0	75	3.5
39	21.7	57	2.6	87	17.1
40	3.4	59	2.0	88	1.0
41	3.1				

Example 85   2-Ethylhexanoic acid ($C_8H_{16}O_2$)

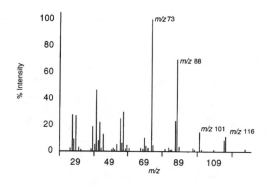

m/z	%	m/z	%	m/z	%
26	2.3	52	1.0	81	1.8
27	27.3	53	5.2	83	2.6
28	9.2	55	24.4	84	1.2
29	26.7	56	6.1	85	1.2
31	2.8	57	29.5	87	23.0
32	1.0	58	1.7	88	69.9
38	1.9	59	4.6	89	3.4
39	18.6	60	2.2	97	2.1
40	5.3	67	2.0	98	1.4
41	46.3	68	1.6	101	14.8
42	7.9	69	9.8	102	1.3
43	21.7	70	3.7	109	1.1
44	2.2	71	2.2	115	8.4
45	12.8	73	100.0	116	11.2
50	1.2	74	4.6	127	1.9
51	2.0				

Example 86   Benzoic acid (C₇H₆O₂)

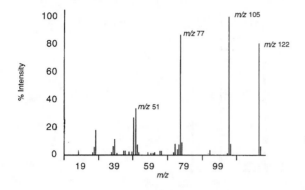

m/z	%	m/z	%	m/z	%
18	3.0	50	27.1	74	7.9
26	1.7	51	33.3	75	3.9
27	5.8	52	7.1	76	7.1
28	17.9	53	1.9	77	86.4
37	1.7	58	1.7	78	9.2
38	6.2	61	1.2	94	3.5
39	11.2	62	1.1	104	1.3
40	1.4	63	1.9	105	100.0
44	2.6	65	3.0	106	7.6
45	2.9	66	2.7	122	80.3
47	2.0	73	1.7	123	6.2
49	2.2				

Example 87   4-Methylbenzoic acid (C₈H₈O₂)

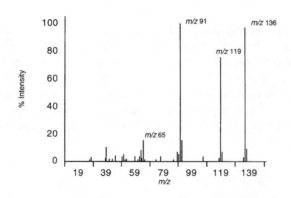

m/z	%	m/z	%	m/z	%
27	1.4	59	3.5	90	5.3
28	2.7	61	1.3	91	100.0
38	2.0	62	3.3	92	14.9
39	9.8	63	7.9	107	3.5
41	1.8	64	2.0	118	2.0
43	1.8	65	15.3	119	75.6
45	3.8	66	1.1	120	6.8
50	3.5	74	1.1	135	2.6
51	4.8	77	3.1	136	97.3
52	1.1	86	1.0	137	8.8
53	1.6	89	6.5		

Example 88   2-Methylbenzoic acid ($C_8H_8O_2$)

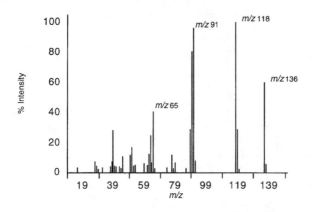

m/z	%	m/z	%	m/z	%
1 6	3.1	5 0	12.0	7 8	3.0
2 7	7.4	5 1	16.8	7 9	6.8
2 8	4.6	5 2	4.7	8 6	3.0
2 9	2.3	5 3	4.8	8 9	28.7
3 2	3.2	5 9	6.3	9 0	80.5
3 7	3.8	6 1	4.8	9 1	96.2
3 8	7.4	6 2	12.4	9 2	7.8
3 9	27.8	6 3	24.8	118	100.0
4 0	4.7	6 4	6.6	119	28.3
4 1	4.2	6 5	40.3	120	2.1
4 3	4.0	6 6	2.8	136	59.9
4 4	2.6	7 4	3.6	137	5.5
4 5	10.5	7 7	11.6		

Ordinarily, the molecular ions of esters and acids will have low intensities, but in many cases these ions will be detectable. Aromatic esters and acids can have very intense molecular ions. For the esters, we can write three general structures (Figure 9.1) where R and Ar represent the alkyl or aromatic group. Ordinary α-cleavage would be expected to produce six fragment ions (and six corresponding neutrals) by cleavage of the indicated bonds in Figure 9.2, which are shown for an aliphatic ester. These expected fragment ions would be R^+, RCO^+, RCO_2^+, R'^+, $R'O^+$ and $R'OCO^+$. Esters containing aromatic

$$R - \overset{\overset{\displaystyle O}{\|}}{C} - O - R$$

$$Ar - \overset{\overset{\displaystyle O}{\|}}{C} - O - R$$

$$R - \overset{\overset{\displaystyle O}{\|}}{C} - O - Ar$$

Figure 9.1

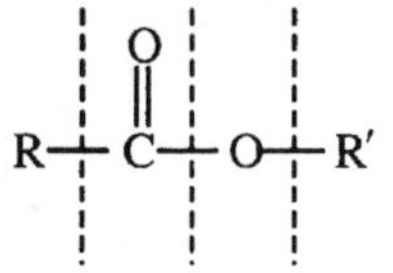

$$R \!-\!\!\!\!\underset{}{\overset{}{|}}\!\!\!\!- \overset{\displaystyle O}{\underset{}{\overset{\|}{C}}} \!-\!\!\!\!\underset{}{\overset{}{|}}\!\!\!\!- O \!-\!\!\!\!\underset{}{\overset{}{|}}\!\!\!\!- R'$$

Figure 9.2

groups would yield the corresponding aromatic ions. In practice, the spectra of esters tend to be dominated by the RCO^+, $ArCO^+$ and the McLafferty rearrangement peaks. The other fragment ions may be present but in most cases will not be particularly significant. In the majority of esters, fragmentation normally occurs between the carbonyl carbon and the oxygen bridge, with the charge retained on the carbonyl fragment.

In addition to the α-cleavage and the McLafferty rearrangement, there are two other processes that are important in ester fragmentation. The first process is γ-cleavage in the acid chain of the ester (Figure 9.3) to produce an ion series starting with m/z 87. These ions have been shown to have the structure indicated in Figure 9.4 and are resonance stabilized. Other ions in the series occur at m/z 101, 115, etc. corresponding to ethyl, propyl and higher esters.

The second mechanism, which is related to the McLafferty rearrangement, is a double hydrogen transfer (DHT). This mechanism occurs in the alcohol chain for ethyl and higher esters and can be viewed as a two step process. In the first step, hydrogen from the adjoining carbon protonates the oxygen bridge. This is followed by hydrogen transfer from a γ-atom and a McLafferty style rearrangement. This mechanism produces ions 1 amu higher than those given by the McLafferty rearrangement. It should also be noted that, whereas the McLafferty rearrangement produces OE ions, the DHT mechanism produces EE ions. In some cases, the peaks produced by the DHT are more intense than those produced by the McLafferty rearrangement. This mechanism is shown in Figure 9.5.

Although we have examined these reactions for aliphatic esters, similar processes occur in aromatic esters. *Ortho*-substituted esters can lose ROH instead of RO radical to produce OE ions. We have already looked at the *ortho*-effect in some detail in Chapter 6 so one example should suffice. For methyl 2-methylbenzoate, hydrogen transfers from the

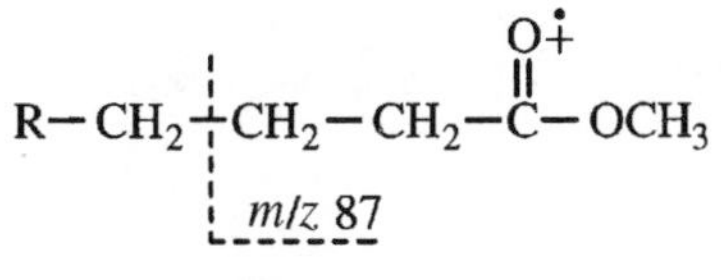

$$R - CH_2 \!-\!\!\!\!\underset{}{\overset{}{|}}\!\!\!\!- CH_2 - CH_2 - \overset{\displaystyle O\overset{\cdot}{+}}{\overset{\|}{C}} - OCH_3$$

m/z 87

Figure 9.3

$$CH_2 \!=\! CH - \overset{\displaystyle \overset{\oplus}{O}H}{\overset{\|}{C}} - OCH_3$$

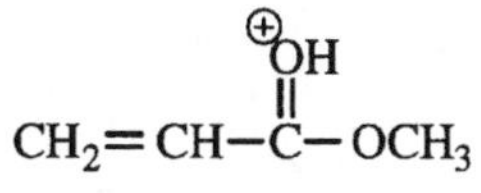

Figure 9.4

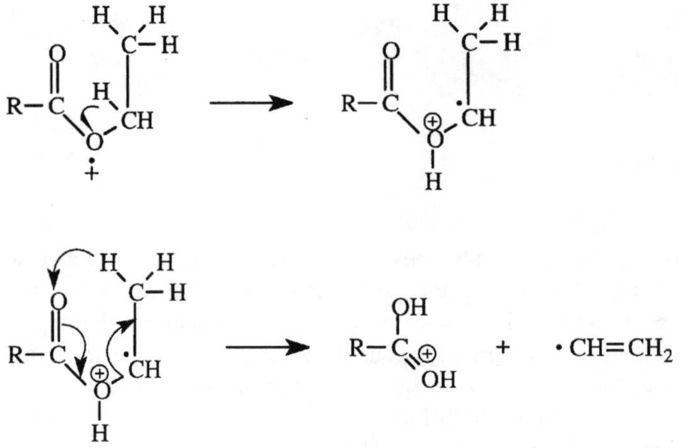

Figure 9.5

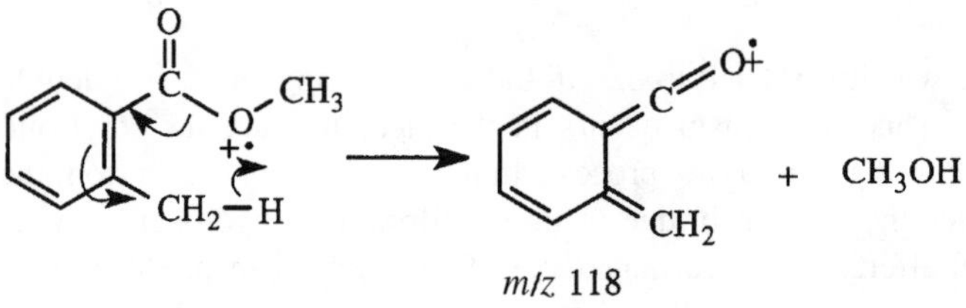

Figure 9.6

m/z 118

methyl group on the benzene ring to the oxygen bridge, producing an OE ion at *m/z* 118 with the elimination of a molecule of methanol (Figure 9.6).

When the aromatic ring is part of the alcohol, rearrangements can occur to eliminate a ketene molecule. In this case, the carboxylic acid must have at least two carbons. In the benzyl esters, this produces a characteristic fragment at *m/z* 108 (Figure 9.7).

Aryl esters such as phenyl ethanoate can also rearrange through a six-member ring transition state, eliminating a ketene molecule and producing a characteristic peak at *m/z* 94, which is often the base peak. This reaction is shown in Figure 9.8.

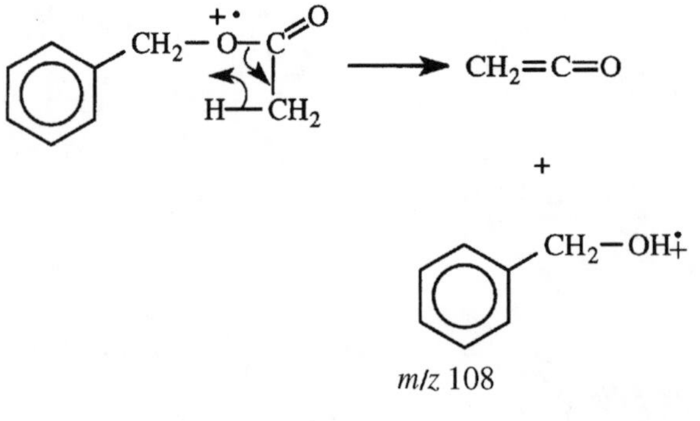

$$CH_2 = C = O$$

m/z 108

Figure 9.7

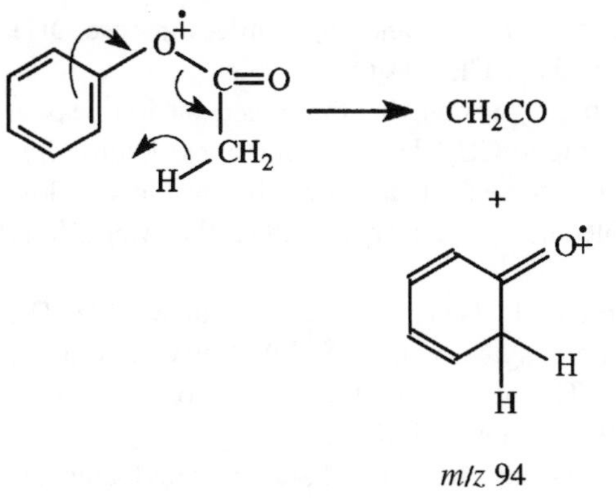

m/z 94

Figure 9.8

Fragmentations of di- and polyesters are beyond the scope of this manual, with one exception. The dialkyl phthalates, which are commonly used plasticizers, are characterized by intense lines at *m/z* 149. This base peak results from a three step process involving loss of RO radical from one of the esters, formation of a five-membered ring and elimination of an alkene. This mechanism is shown in Figure 9.9.

Examples 71 through 79 illustrate most of the points discussed concerning ester spectra. In the spectrum of methyl ethanoate (Example 71) we see a reasonably intense molecular ion at *m/z* 74. The line at *m/z* 59 is due to loss of methyl radical, and the base peak at *m/z* 43 corresponds to CH_3CO^+. In the spectrum of ethyl methanoate (Example 72) we see the molecular ion at *m/z* 74 and the base peak at *m/z* 31 (corresponding to

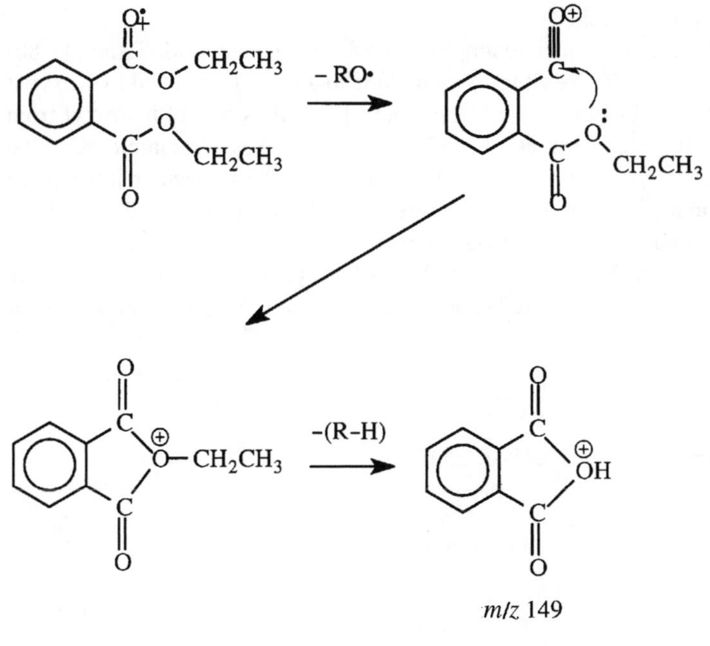

m/z 149

Figure 9.9

HCO^+). We also see a weak McLafferty peak at m/z 56 and an even less intense DHT peak at m/z 57. The peak at m/z 45 corresponds to $CH_3CH_2O^+$.

The spectrum of methyl propanoate (Example 73) contains the molecular ion peak at m/z 88, an intense line at m/z 57 (corresponding to $CH_3CH_2CO^+$) and a less intense line at m/z 59 (corresponding to CH_3OCO^+). The base peak at m/z 29 can be attributed to loss of CO from m/z 57. Notice that this molecule can undergo neither the McLafferty rearrangement (no γ-H) nor the DHT.

The molecular ion in methyl butanoate (Example 74) is very weak but detectable. The line at m/z 87 corresponds to $C_3H_7CO_2^+$. There is a very intense McLafferty peak at m/z 74, while m/z 71 corresponds to $C_3H_7CO^+$. The base peak at m/z 43 is consistent with loss of CO from m/z 71. The line at m/z 59 represents CH_3OCO^+.

The spectra of higher molecular weight esters, such as methyl 2,4,6-trimethyloctanoate (Example 75), exhibit a considerable amount of hydrocarbon character. In this spectrum, the McLafferty rearrangement peak is the base peak at m/z 88. The line at m/z 169 corresponds to loss of $CH_3O^{\cdot}$ from the parent. Other reasonably intense lines such as m/z 29 and m/z 101 represent fragmentation of the hydrocarbon chain.

Ethyl decanoate (Example 76) exhibits both the McLafferty peak at m/z 88 and the DHT peak at m/z 89. The McLafferty peak is at m/z 74 in methyl undecanoate (Example 77). Both of these spectra also exhibit the expected hydrocarbon patterns.

The spectrum of methyl 3-methylbenzoate (Example 78) exhibits characteristic fragmentations for benzoic acid derivatives, including loss of 31 ($CH_3O^{\cdot}$) from the parent and formation of the tropylium ion at m/z 91 by loss of CO from m/z 119. The line at m/z 119 is missing in the spectrum of methyl phenylethanoate (Example 79), which loses CH_3OCO radical to form the tropylium ion directly.

Except for relatively low molecular weight members, the spectra of aliphatic acids exhibit weak molecular ions. The most common ion fragment is $COOH^+$ at m/z 45. Once the acid side chain exceeds three carbons, the McLafferty peak at m/z 60 becomes very intense and is often the base peak.

Typical acid spectra are shown in Examples 80–88. Methanoic acid (Example 80) contains the molecular ion at m/z 46. Other lines include loss of hydrogen and hydroxyl at m/z 45 and 29 respectively. Propanoic acid (Example 81) exhibits loss of hydroxyl from the parent producing the fragment at m/z 57. In butanoic acid (Example 82), the McLafferty rearrangement peak at m/z 60 is the base peak. The molecular ion is very weak (2.5%). Hexanoic acid (Example 83) exhibits the McLafferty peak, and other lines at m/z 73 and 87 correspond to hydrocarbon fragments.

In 2-methylpentanoic acid (Example 84), the McLafferty peak occurs at m/z 74 owing to substitution on the β-carbon. In 2-ethylhexanoic acid (Example 85), this rearrangement

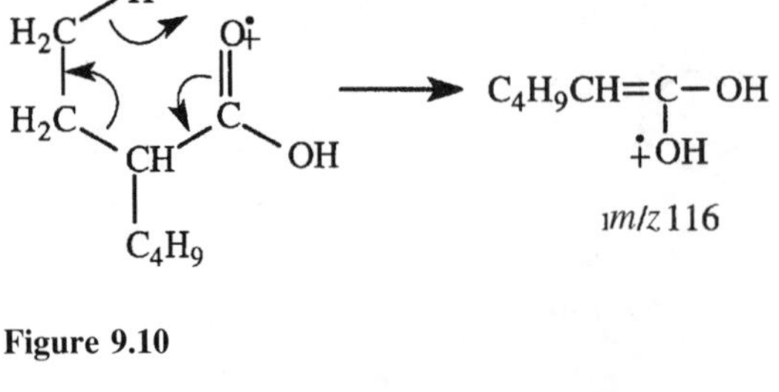

Figure 9.10

peak is present at m/z 88. A second McLafferty peak occurs at m/z 116. This is because the ethyl group substituted at the β-carbon can participate in the McLafferty rearrangement in competition with the longer side chain. This reaction is shown in Figure 9.10.

Aromatic acids tend to produce intense molecular ions, and fragment by the sequential loss of OH and CO. Benzoic acid (Example 86) exhibits these characteristics very clearly. Methylbenzoic acids exhibit the same losses, producing tropylium ions. When the methyl group is *ortho*, loss of water produces an OE ion at m/z 118. Subsequent loss of CO produces the OE line at m/z 90. (Examples 87 and 88).

10 Nitrogen-Containing Compounds

Example 89   2-Propanamine (C_3H_9N)

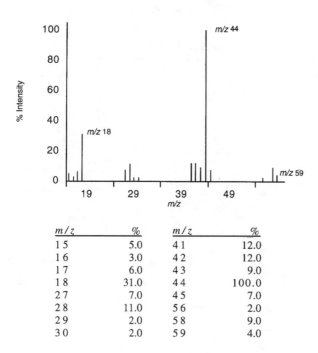

m/z	%	m/z	%
1 5	5.0	4 1	12.0
1 6	3.0	4 2	12.0
1 7	6.0	4 3	9.0
1 8	31.0	4 4	100.0
2 7	7.0	4 5	7.0
2 8	11.0	5 6	2.0
2 9	2.0	5 8	9.0
3 0	2.0	5 9	4.0

Example 90   2-Butanamine (C$_4$H$_{11}$N)

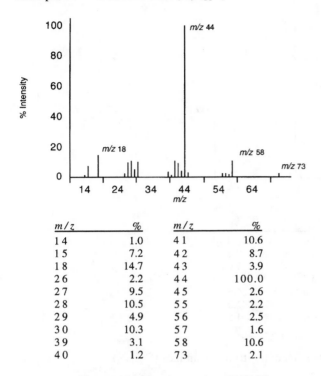

m/z	%	m/z	%
1 4	1.0	4 1	10.6
1 5	7.2	4 2	8.7
1 8	14.7	4 3	3.9
2 6	2.2	4 4	100.0
2 7	9.5	4 5	2.6
2 8	10.5	5 5	2.2
2 9	4.9	5 6	2.5
3 0	10.3	5 7	1.6
3 9	3.1	5 8	10.6
4 0	1.2	7 3	2.1

Example 91   N-Methylmethanamine (C$_2$H$_7$N)

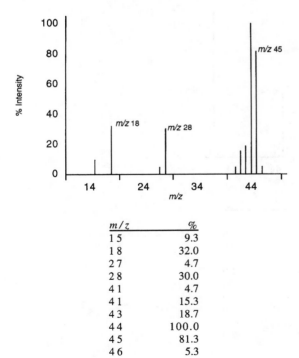

m/z	%
1 5	9.3
1 8	32.0
2 7	4.7
2 8	30.0
4 1	4.7
4 1	15.3
4 3	18.7
4 4	100.0
4 5	81.3
4 6	5.3

Example 92   *N*-methylethanamine (C$_3$H$_9$N)

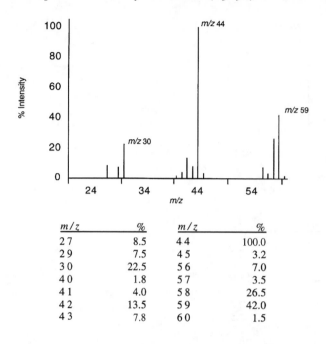

m/z	%	*m/z*	%
2 7	8.5	4 4	100.0
2 9	7.5	4 5	3.2
3 0	22.5	5 6	7.0
4 0	1.8	5 7	3.5
4 1	4.0	5 8	26.5
4 2	13.5	5 9	42.0
4 3	7.8	6 0	1.5

Example 93   *N,N*-Dimethylmethanamine (trimethylamine) (C$_3$H$_9$N)

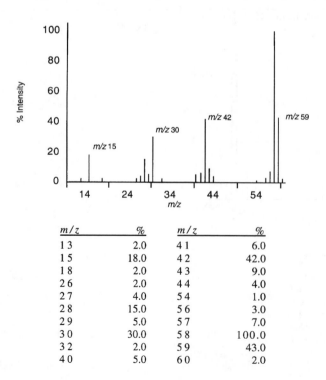

m/z	%	*m/z*	%
1 3	2.0	4 1	6.0
1 5	18.0	4 2	42.0
1 8	2.0	4 3	9.0
2 6	2.0	4 4	4.0
2 7	4.0	5 4	1.0
2 8	15.0	5 6	3.0
2 9	5.0	5 7	7.0
3 0	30.0	5 8	100.0
3 2	2.0	5 9	43.0
4 0	5.0	6 0	2.0

Example 94   *N*-Ethylethanamine (C$_4$H$_{11}$N)

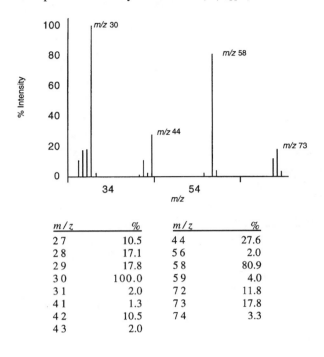

m/z	%	m/z	%
27	10.5	44	27.6
28	17.1	56	2.0
29	17.8	58	80.9
30	100.0	59	4.0
31	2.0	72	11.8
41	1.3	73	17.8
42	10.5	74	3.3
43	2.0		

Example 95   *N*,*N*-Dimethylethanamine (C$_4$H$_{11}$N)

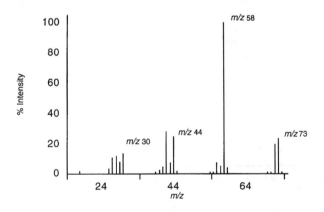

m/z	%	m/z	%	m/z	%
18	1.7	41	4.6	57	5.0
26	3.4	42	27.8	58	100.0
27	10.6	43	7.3	59	4.0
28	11.6	44	24.8	70	1.4
29	8.0	45	1.8	71	1.0
30	13.3	54	1.3	72	19.3
39	1.2	55	1.0	73	23.3
40	2.1	56	7.2	74	1.1

Example 96 *N*-Methyl-1-pentanamine (C$_6$H$_{15}$N)

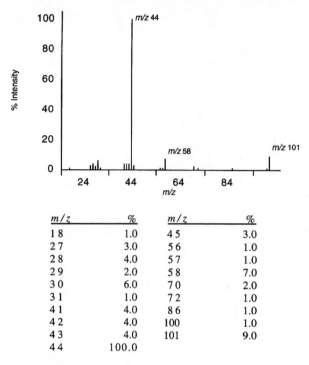

m/z	%	m/z	%
1 8	1.0	4 5	3.0
2 7	3.0	5 6	1.0
2 8	4.0	5 7	1.0
2 9	2.0	5 8	7.0
3 0	6.0	7 0	2.0
3 1	1.0	7 2	1.0
4 1	4.0	8 6	1.0
4 2	4.0	100	1.0
4 3	4.0	101	9.0
4 4	100.0		

Example 97 3,3-Dimethyl-2-butanamine (C$_6$H$_{15}$N)

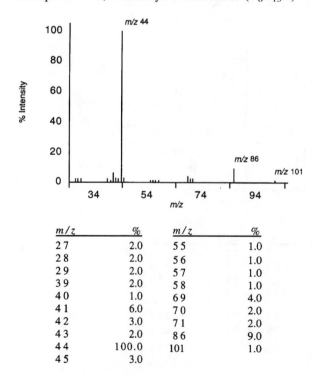

m/z	%	m/z	%
2 7	2.0	5 5	1.0
2 8	2.0	5 6	1.0
2 9	2.0	5 7	1.0
3 9	2.0	5 8	1.0
4 0	1.0	6 9	4.0
4 1	6.0	7 0	2.0
4 2	3.0	7 1	2.0
4 3	2.0	8 6	9.0
4 4	100.0	101	1.0
4 5	3.0		

Example 98   2-Methylaniline (C$_7$H$_9$N)

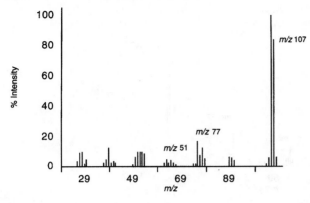

m/z	%	m/z	%	m/z	%
26	3.3	51	9.4	77	17.0
27	9.1	52	9.6	78	7.1
28	9.4	53	9.6	79	12.5
29	1.5	54	8.6	80	5.3
30	4.3	62	2.0	89	6.0
37	2.2	63	4.6	90	5.5
38	4.7	64	2.0	91	3.8
39	12.3	65	4.0	104	1.8
40	2.3	66	2.0	105	5.4
41	3.5	67	1.0	106	100.0
42	2.4	74	1.7	107	83.1
49	1.0	76	1.8	108	6.2
50	6.2				

Example 99   Aniline (C$_6$H$_7$N)

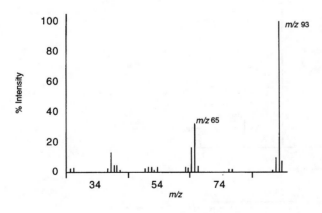

m/z	%	m/z	%	m/z	%
26	2.0	51	3.2	67	3.9
27	3.0	52	3.2	77	1.5
38	2.4	53	1.3	78	1.5
39	13.0	54	3.4	91	1.4
40	4.5	63	3.1	92	9.7
41	4.6	64	2.6	93	100.0
42	1.4	65	16.4	94	7.1
50	2.3	66	32.0		

112

Example 100   N-Methylaniline (C$_7$H$_9$N)

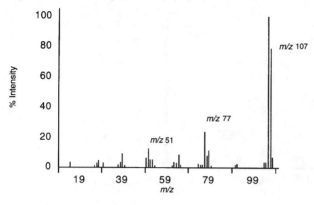

m/z	%	m/z	%	m/z	%
15	3.5	52	5.1	77	23.4
26	1.1	53	5.2	78	8.0
27	2.7	54	1.1	79	11.3
28	4.4	62	1.4	80	1.2
30	3.0	63	3.6	91	1.5
37	1.7	64	2.9	92	2.4
38	3.5	65	8.6	104	3.6
39	9.0	66	1.6	105	3.4
40	1.1	74	2.1	106	100.0
50	6.3	75	1.5	107	79.0
51	12.3	76	1.5	108	6.8

Example 101   Butanamide (C$_4$H$_9$NO)

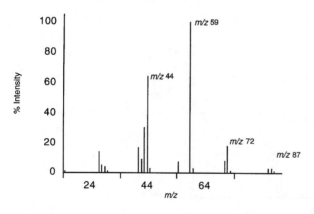

m/z	%	m/z	%
15	1.0	55	7.0
27	14.0	59	100.0
28	5.0	60	3.0
29	4.0	71	8.0
30	1.0	72	18.0
41	17.0	73	1.0
42	9.0	86	3.0
43	30.0	87	3.0
44	64.0	88	1.0
45	3.0		

Example 102   Benzamide (C$_7$H$_7$NO)

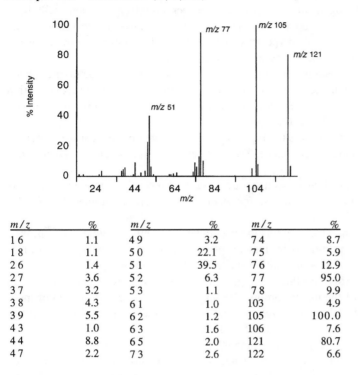

m / z	%	m / z	%	m / z	%
16	1.1	49	3.2	74	8.7
18	1.1	50	22.1	75	5.9
26	1.4	51	39.5	76	12.9
27	3.6	52	6.3	77	95.0
37	3.2	53	1.1	78	9.9
38	4.3	61	1.0	103	4.9
39	5.5	62	1.2	105	100.0
43	1.0	63	1.6	106	7.6
44	8.8	65	2.0	121	80.7
47	2.2	73	2.6	122	6.6

Example 103   N-Ethylbenzamide (C$_9$H$_{11}$NO)

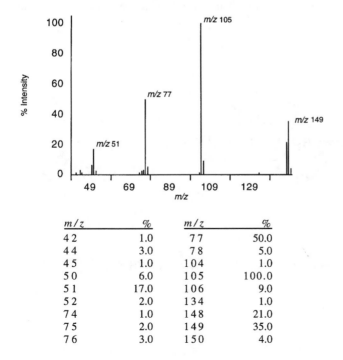

m / z	%	m / z	%
42	1.0	77	50.0
44	3.0	78	5.0
45	1.0	104	1.0
50	6.0	105	100.0
51	17.0	106	9.0
52	2.0	134	1.0
74	1.0	148	21.0
75	2.0	149	35.0
76	3.0	150	4.0

Probably the most important point for the student to remember is that compounds containing a single nitrogen atom, or an odd number of nitrogen atoms, will have molecular ions occurring at odd masses. This is the application of the nitrogen rule discussed earlier. The spectra of compounds containing an odd number of nitrogen atoms are dominated by fragments occurring at even mass values.

The α-cleavage mechanism is very important in understanding the spectra of aliphatic amines. There are two other mechanisms that are important. The first is a ring formation reaction producing a series of even-electron fragments at m/z 58, 72, 86, 100 and 114. This reaction is shown in Figure 10.1.

The second mechanism involves the transfer of a β-hydrogen to the positively charged nitrogen and the elimination of an alkene. This transfer takes place after α-cleavage and is shown in Figure 10.2.

The molecular ions of aliphatic amines are usually observable, although weak. The molecular ions are usually more intense for secondary and tertiary amines than for the corresponding primary amines.

Example 89 (2-propanamine) exhibits the molecular ion at m/z 59 and the base peak at m/z 44. The base peak is formed by loss of either methyl group. The spectrum of 2-butanamine (Example 90) is superficially very similar to that of 2-propanamine, with the base peak at m/z 44 (loss of the ethyl group) and a less intense peak at m/z 58 corresponding to loss of the methyl group. The major differences in the spectra of these two compounds are the absence of m/z 59 and the presence of an easily overlooked molecular ion at m/z 73 in the spectrum of 2-butanamine.

The spectrum of N-methylmethanamine (Example 91) exhibits a very intense molecular ion at m/z 45. The base peak at m/z 44 is due to loss of hydrogen, and the peak at m/z 28 is consistent with $CH=NH^+$. N-Methylethanamine (Example 92) also exhibits a reasonably intense molecular ion. The base peak results from α-cleavage. This spectrum also contains a line at m/z 30 which is characteristic for many amines and corresponds to $CH_2=NH_2^+$.

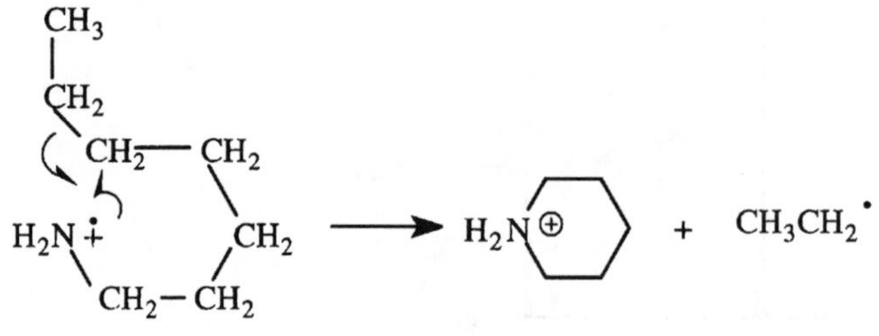

Figure 10.1

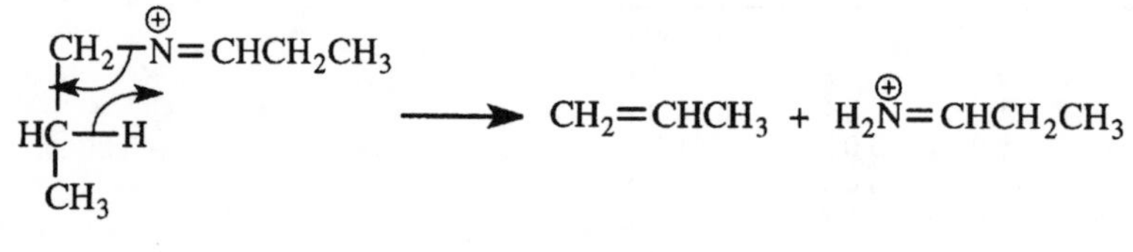

Figure 10.2

N,N-Dimethylmethanamine (Example 93) exhibits a very intense molecular ion at *m/z* 59. The line at *m/z* 42 is consistent with the loss of ($CH_3 + H_2$). *N*-Ethylethanamine (Example 94) exhibits the expected α-cleavage line at *m/z* 58. In this spectrum, there are two β-cleavage peaks. The base peak is consistent with loss of $CH_2=CH_2$ from *m/z* 58. The second β-cleavage peak occurs at *m/z* 44 and is produced from *m/z* 72. *N,N*-Dimethylethanamine (Example 95) exhibits a reasonably intense molecular ion and a base peak at *m/z* 58 due to α-cleavage.

N-Methyl-1-pentanamine (Example 96) exhibits the molecular ion at *m/z* 101 and a base peak at *m/z* 44 due to α-cleavage. The peak at *m/z* 58 is consistent with the ring formation reaction described above. In the spectrum of 3,3-dimethyl-2-butanamine (Example 97) we can observe lines at *m/z* 86 and 44, with loss of the largest alkyl predominating.

Aromatic amines, such as 2-methylaniline (Example 98), exhibit the same increase in stability of the molecular ion that we have observed in other aromatic compounds. The intense line at *m/z* 106 is an azatropylium ion (Figure 10.3) which loses HCN to form the important OE ion at *m/z* 79. Similar losses of HCN from the molecular ion and from (M − H) account for the lines at *m/z* 66 and 65 in the spectrum of aniline (Example 99). *N*-Methylaniline (Example 100) produces a spectrum which is almost indistinguishable from that of 2-methylaniline.

The fragmentation patterns seen in the spectra of amides are similar to those seen in the spectra of acids and esters. The predominant mechanism is α-cleavage, with the alkyl group leaving as a radical and the charge being retained on the $CONH_2$ ion. In an amide, there are two locations where ionization generally occurs; the oxygen atom or the nitrogen atom. In either case the same fragment ion is produced (Figure 10.4). The ion produced is resonance stabilized. Substitution on the nitrogen atom produces an ion series at *m/z* 44, 58, 72 etc.

Acylium ions (RCO^+) tend to be weak in the spectra of aliphatic amides. In aromatic amides, these acylium ion peaks can have considerable intensity. This is due to the ability

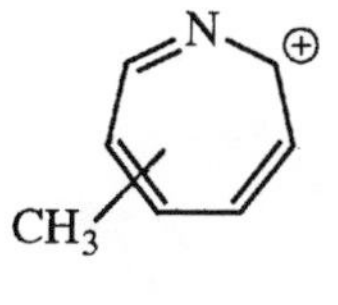

Figure 10.3

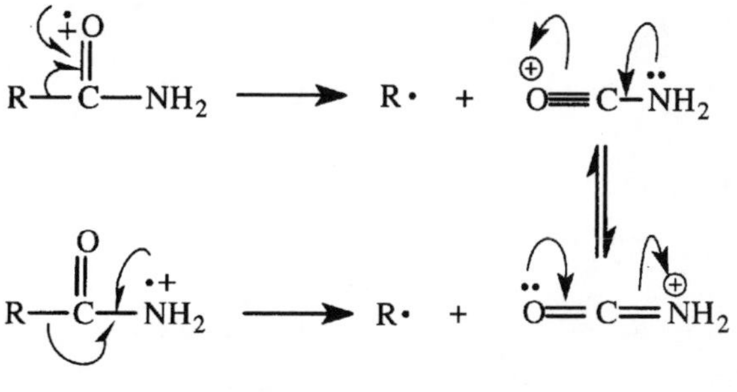

Figure 10.4

of aromatic groups to donate electrons from their π-orbital systems and stabilize the positive charge by resonance.

The McLafferty rearrangement can occur if the acid chain is longer than three carbons. Secondary and tertiary amides will produce significant amounts of substituted McLafferty ions. Cleavage between the β- and γ-carbons in the acid chain is another important reaction mechanism. In primary amides, this produces a fragment at m/z 72 with the formula $^{+}CH_2CH_2CONH_2$. This ion can lose ketene ($CH_2=C=O$) to produce $CH_2=NH_2^{+}$ at m/z 30. In secondary and tertiary amides, β-cleavage can also occur in the N-alkyl chains.

Aromatic amides such as acetanilide ($C_6H_5NHCOCH_3$) initially fragment by losing a ketene (or substituted ketene), producing the base peak of the corresponding aniline. The spectra of these aromatic amides closely resemble the spectra of the corresponding anilines.

In the spectrum of butanamide (Example 101) we see a weak molecular ion at m/z 87 (about 3.0%). We also see lines at m/z 72 (cleavage between the β- and γ-carbons), m/z 59 (the McLafferty rearrangement peak), and m/z 44 (loss of the alkyl group). The spectrum of benzamide (Example 102) exhibits the molecular ion at m/z 121. Other peaks include m/z 105 (loss of NH_2), m/z 77 (loss of CO from m/z 105) and m/z 51 (loss of C_2H_2 from m/z 77). The student will recognize the pair of peaks at m/z 77 and 51 as characteristic for monosubstituted benzenes. Finally, the spectrum of N-ethylbenzamide (Example 103) is very similar to that of benzamide. The only significant differences are the molecular ion at m/z 149, and the loss of C_2H_5NH from the molecular ion to form the acylium ion at m/z 105.

11 Thiols and Thioethers

Example 104   1-Propanethiol (C$_3$H$_8$S)

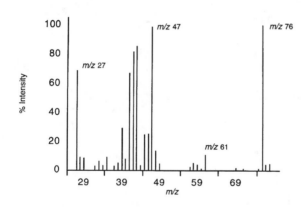

m/z	%	m/z	%	m/z	%
27	68.6	41	66.8	58	5.1
28	9.0	42	81.4	59	4.1
29	8.6	43	85.6	60	1.1
32	2.9	44	3.4	61	10.9
33	5.9	45	24.7	69	1.7
34	3.5	46	25.4	71	1.4
35	8.8	47	99.1	75	1.3
37	2.9	48	13.4	76	100.0
38	5.3	49	4.7	77	3.9
39	28.9	57	2.3	78	4.5
40	7.7				

Example 105   2-Propanethiol (C_3H_8S)

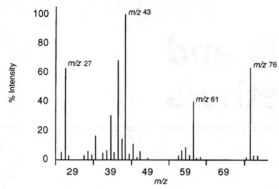

m/z	%	m/z	%	m/z	%
2 6	4.8	4 0	5.3	5 8	6.1
2 7	62.6	4 1	68.0	5 9	8.4
2 8	2.8	4 2	14.0	6 0	2.7
3 2	2.6	4 3	100.0	6 1	39.4
3 3	5.8	4 4	3.7	6 2	1.2
3 4	3.1	4 5	10.4	6 3	1.7
3 5	16.3	4 6	1.5	7 5	1.5
3 7	4.4	4 7	5.5	7 6	63.4
3 8	6.2	4 9	1.1	7 7	2.6
3 9	29.9	5 7	2.9	7 8	2.8

Example 106   2-Butanethiol ($C_4H_{10}S$)

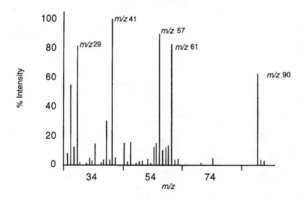

m/z	%	m/z	%	m/z	%
2 6	7.6	4 1	100.0	5 7	89.9
2 7	54.8	4 2	4.8	5 8	10.3
2 8	12.1	4 5	15.3	5 9	11.6
2 9	81.3	4 6	2.3	6 0	13.3
3 0	1.8	4 7	15.6	6 1	82.6
3 2	1.3	4 9	1.4	6 2	3.2
3 3	4.7	5 0	2.4	6 3	3.7
3 4	2.9	5 1	2.9	7 1	1.3
3 5	14.3	5 3	3.9	7 5	4.5
3 7	1.8	5 4	1.0	9 0	62.5
3 8	3.1	5 5	12.1	9 1	3.2
3 9	30.4	5 6	15.0	9 2	2.8
4 0	3.4				

Example 107   1-Pentanethiol ($C_5H_{12}S$)

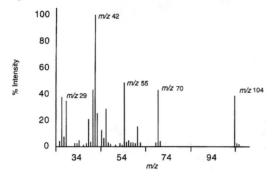

m/z	%	m/z	%	m/z	%
26	4.2	42	100.0	57	4.3
27	37.3	43	25.3	58	2.6
28	7.0	45	12.1	59	3.0
29	34.5	46	6.3	60	2.5
33	2.0	47	28.3	61	14.9
34	2.1	48	2.7	62	2.9
35	4.6	49	1.8	69	2.6
37	1.1	51	1.1	70	42.9
38	2.1	53	2.0	71	4.1
39	20.6	54	1.0	104	38.5
40	3.5	55	48.8	105	2.5
41	43.0	56	3.6	106	1.7

Example 108   2-Methyl-3-pentanethiol ($C_6H_{14}S$)

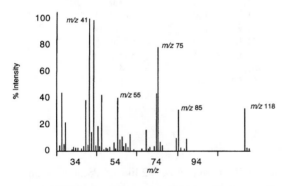

m/z	%	m/z	%	m/z	%
26	3.7	46	3.3	69	15.5
27	43.8	47	42.1	70	1.6
28	4.8	48	1.2	71	2.7
29	21.5	49	2.3	73	3.4
32	1.1	50	1.7	74	43.1
33	3.0	51	2.7	75	78.1
34	2.1	53	6.2	76	6.9
35	2.0	54	1.6	77	4.0
37	1.5	55	39.4	84	9.3
38	3.4	56	8.2	85	30.9
39	38.2	57	10.8	86	2.1
40	4.4	58	3.5	88	1.5
41	100.0	59	5.5	89	8.9
42	14.1	60	2.9	118	31.8
43	99.1	61	12.5	119	2.3
44	3.7	63	1.3	120	1.5
45	18.4	67	1.6		

Example 109   1-Octanethiol (C$_8$H$_{18}$S)

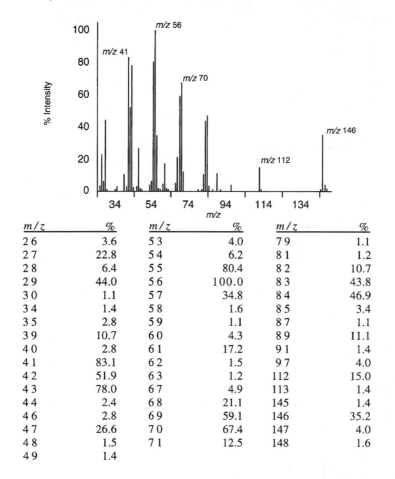

m/z	%	m/z	%	m/z	%
26	3.6	53	4.0	79	1.1
27	22.8	54	6.2	81	1.2
28	6.4	55	80.4	82	10.7
29	44.0	56	100.0	83	43.8
30	1.1	57	34.8	84	46.9
34	1.4	58	1.6	85	3.4
35	2.8	59	1.1	87	1.1
39	10.7	60	4.3	89	11.1
40	2.8	61	17.2	91	1.4
41	83.1	62	1.5	97	4.0
42	51.9	63	1.2	112	15.0
43	78.0	67	4.9	113	1.4
44	2.4	68	21.1	145	1.4
46	2.8	69	59.1	146	35.2
47	26.6	70	67.4	147	4.0
48	1.5	71	12.5	148	1.6
49	1.4				

Sulfur-containing compounds will exhibit an obvious A + 2 line of about 4.4% and as a rule will have more intense parent ions than the corresponding oxygen compounds. The same kinds of reactions that we have seen for alcohols and ethers will occur in the sulfur analogs. There are a few different reactions that the sulfur-containing compounds will undergo, and we will consider them when appropriate.

Thiols will exhibit an isotope series at m/z 47, 61, 75, 89 etc. corresponding to the alcohol series at m/z 31, 45, 59, 73 etc. These ions are produced by α-cleavage, with the ion at m/z 47 corresponding to CH$_2$=SH$^+$, and higher members corresponding to various substituents on the carbon. Thiols exhibit (M − 34) and (M − 62) lines corresponding to loss of H$_2$S and (H$_2$S + CH$_2$=CH$_2$) respectively. Hydrocarbon ions formed by cleavage at each point in the alkyl chain will also be found in the spectra of most thiols.

In chains containing five or more carbons, δ-cleavage and cyclization will occur, producing a 5-membered ring at m/z 89. This reaction is shown in Figure 11.1.

Example 110   Benzenethiol (C_6H_6S)

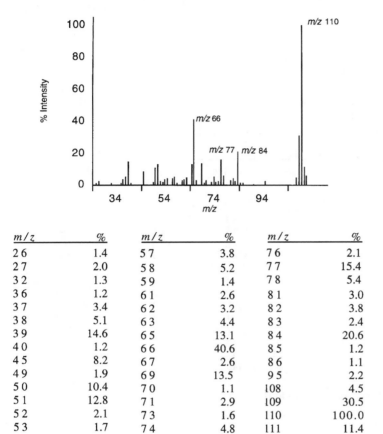

m/z	%	m/z	%	m/z	%
2 6	1.4	5 7	3.8	7 6	2.1
2 7	2.0	5 8	5.2	7 7	15.4
3 2	1.3	5 9	1.4	7 8	5.4
3 6	1.2	6 1	2.6	8 1	3.0
3 7	3.4	6 2	3.2	8 2	3.8
3 8	5.1	6 3	4.4	8 3	2.4
3 9	14.6	6 5	13.1	8 4	20.6
4 0	1.2	6 6	40.6	8 5	1.2
4 5	8.2	6 7	2.6	8 6	1.1
4 9	1.9	6 9	13.5	9 5	2.2
5 0	10.4	7 0	1.1	108	4.5
5 1	12.8	7 1	2.9	109	30.5
5 2	2.1	7 3	1.6	110	100.0
5 3	1.7	7 4	4.8	111	11.4
5 4	3.4	7 5	1.9	112	5.7
5 5	3.9				

In the spectra of thioethers (sulfides) we would expect to see α-cleavage producing $RS^+=CH_2$, with Stevenson's rule indicating that loss of the largest alkyl will be favored. If the remaining R group contains a β-hydrogen, the next step would be transfer of this hydrogen to the sulfur and elimination of an alkene, forming $CH_2=SH^+$. These reactions are similar to those we have seen in ethers.

Figure 11.1

Example 111   1-(Butylthio)butane (di-n-butyl sulfide) (C$_8$H$_{18}$S)

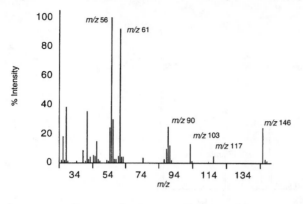

m/z	%	m/z	%	m/z	%
2 6	1.6	4 7	14.5	6 3	4.2
2 7	17.9	4 8	2.4	7 5	3.5
2 8	1.9	4 9	1.3	8 8	2.2
2 9	38.0	5 3	1.8	8 9	9.4
3 0	1.1	5 4	1.4	9 0	24.6
3 5	1.3	5 5	23.8	9 1	12.0
3 9	8.5	5 6	100.0	9 2	1.5
4 0	1.2	5 7	29.6	103	12.8
4 1	35.4	5 8	2.2	104	1.4
4 2	2.3	5 9	2.0	117	4.3
4 3	3.7	6 0	4.5	146	24.3
4 5	5.1	6 1	92.2	147	2.5
4 6	4.4	6 2	4.2	148	1.2

Example 112   Benzyl methyl sulfide (C$_8$H$_{10}$S)

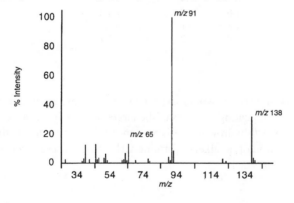

m/z	%	m/z	%	m/z	%
2 7	2.3	5 2	1.6	8 9	3.7
3 7	1.0	6 1	1.6	9 0	1.6
3 8	2.7	6 2	2.4	9 1	100.0
3 9	12.2	6 3	6.5	9 2	8.3
4 1	2.1	6 4	1.8	121	2.9
4 5	13.0	6 5	13.0	123	1.4
4 6	2.1	6 9	1.8	138	31.6
4 7	3.1	7 7	2.9	139	3.2
5 0	3.3	7 8	1.2	140	1.6
5 1	6.4				

Example 113   Dimethyl sulfoxide (C$_2$H$_6$SO)

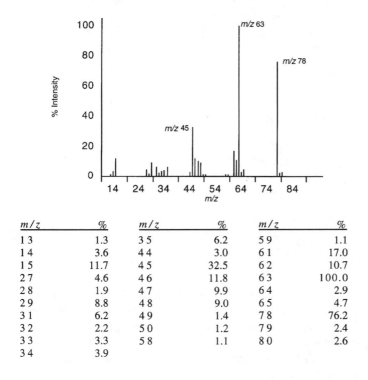

m/z	%	m/z	%	m/z	%
1 3	1.3	3 5	6.2	5 9	1.1
1 4	3.6	4 4	3.0	6 1	17.0
1 5	11.7	4 5	32.5	6 2	10.7
2 7	4.6	4 6	11.8	6 3	100.0
2 8	1.9	4 7	9.9	6 4	2.9
2 9	8.8	4 8	9.0	6 5	4.7
3 1	6.2	4 9	1.4	7 8	76.2
3 2	2.2	5 0	1.2	7 9	2.4
3 3	3.3	5 8	1.1	8 0	2.6
3 4	3.9				

The spectra of thioethers are somewhat different from those of ordinary ethers. The RS$^+$ ions are more stable than the corresponding RO$^+$ ions, and therefore RS$^+$ ions will have relatively high intensities in the series m/z 47, 61, 75 etc. A thioether of the formula R−S−R′ will exhibit RS$^+$ and R′S$^+$ ions of reasonable intensity. Another difference is that one of the alkyl groups can transfer a β-hydrogen to the sulfur, producing a fragment 1 amu higher than the RS$^+$ ion (or the R′S$^+$ ion). The ion produced by β-hydrogen transfer is an OE ion, and is very useful in determining the position of the sulfur.

Arylthioethers will generally have intense molecular ions, and will tend to lose the alkyl group, producing ArS$^+$. The ArS$^+$ ion will lose C=S to form the cyclopentadienyl ion. Methyl arylthioethers will tend to lose CH$_2$=S and transfer a hydrogen to the aromatic ring. The molecular ion will also lose RS, producing Ar$^+$. Long chain alkyls will transfer hydrogen to the sulfur-producing ArSH$^+$. These fragmentation pathways are summarized in Figure 11.2 for methylthiobenzene. The student should be aware that the benzene (m/z 78) and tropylium (m/z 91) ions produced by these processes will fragment in the normal fashion, i.e. there will also be a peak at m/z 65 corresponding to loss of C$_2$H$_2$ from the tropylium ion.

1-Propanethiol (Example 104) exhibits an intense molecular ion at m/z 76. The intense line at m/z 47 corresponds to CH$_2$=SH$^+$ and is accompanied by another reasonably

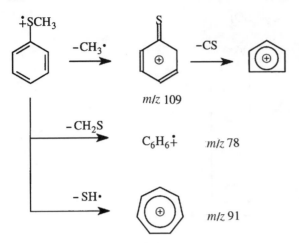

Figure 11.2

intense line at m/z 43 (loss of SH from the molecular ion). The spectrum of 2-propanethiol (Example 105) exhibits a less intense molecular ion than does propanethiol. The line at m/z 61 is an α-cleavage peak with loss of either methyl group. Note that α-cleavage is much more favorable in 2-propanethiol than in 1-propanethiol. The line at m/z 47 is negligible, since it is very difficult to form $CH_2{=}SH^+$ from this molecule, and loss of SH produces the base peak at m/z 43.

The spectrum of 2-butanethiol (Example 106) exhibits the molecular ion at m/z 90. The line at m/z 61 corresponds to CH_3CHSH^+, while the line at m/z 57 corresponds to (M − SH). Lines at m/z 29 and 41 correspond to C_2H_5 and C_3H_5 respectively. 1-Pentanethiol (Example 107) exhibits loss of H_2S (m/z 70) followed by loss of C_2H_4 (m/z 42). Lines at m/z 47 and 61 are part of the ion series described above, while the peak at m/z 55 corresponds to $C_4H_7{}^+$. The spectrum of 2-methyl-3-pentanethiol (Example 108) exhibits considerable hydrocarbon character. The line at m/z 85 indicates loss of SH, and the prominent m/z 75 is an α-cleavage peak.

1-Octanethiol (Example 109) exhibits a reasonably intense molecular ion at m/z 146. Lines at m/z 112 and 84 correspond to loss of H_2S followed by loss of C_2H_4, and the base peak at m/z 56 corresponds to loss of C_2H_4 from m/z 84. The δ-cleavage line appears at m/z 89, although it is of relatively low intensity. The balance of the spectrum is dominated by hydrocarbon fragments.

The spectrum of benzenethiol (Example 110) exhibits loss of hydrogen to form the thiotropylium ion at m/z 109. This ion can presumably lose CS to form the cyclopentadienyl ion at m/z 65. Other lines in the spectrum correspond to M − C_2H_4 (m/z 84), M − SH (m/z 77) and M − CS (m/z 66).

The spectrum of 1-(butylthio)butane (Example 111) exhibits an OE peak at m/z 90. This is the β-hydrogen transfer peak, corresponding to $C_4H_9H^{+\cdot}$. The peak at m/z 61 is an α-cleavage peak formed from m/z 90, while m/z 56 is a hydrocarbon peak (C_4H_8) produced by loss of H_2S from m/z 90. Benzyl methyl sulfide (Example 112) forms the expected tropylium ion at m/z 91 by loss of CH_3S.

Finally, the spectrum of dimethyl sulfoxide (Example 113) is included because this compound is an important industrial solvent. The base peak in this spectrum (m/z 63) corresponds to CH_3SO^+, while m/z 45 corresponds to CHS^+, indicating that significant rearrangements have occurred.

12 Heterocyclic Compounds

Example 114   Pyrrole (C$_4$H$_5$N)

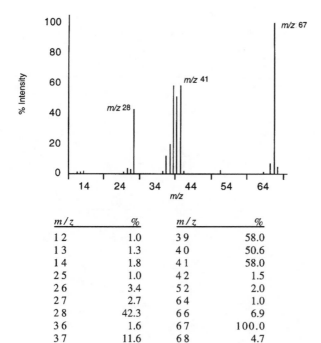

m/z	%	m/z	%
1 2	1.0	3 9	58.0
1 3	1.3	4 0	50.6
1 4	1.8	4 1	58.0
2 5	1.0	4 2	1.5
2 6	3.4	5 2	2.0
2 7	2.7	6 4	1.0
2 8	42.3	6 6	6.9
3 6	1.6	6 7	100.0
3 7	11.6	6 8	4.7
3 8	19.8		

Example 115   Furan (C_4H_4O)

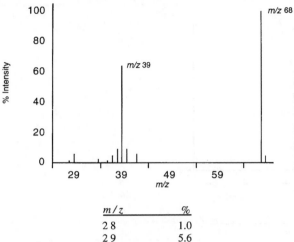

m/z	%
28	1.0
29	5.6
34	2.0
37	4.7
38	8.8
39	63.4
40	8.8
42	5.8
68	100.0
69	4.3

Example 116   Imidazole ($C_3H_4N_2$)

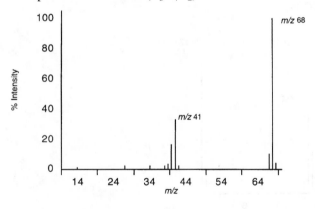

m/z	%
14	1.2
27	2.4
34	2.5
38	2.2
39	3.5
40	16.0
41	33.0
42	2.2
67	10.2
68	100.0
69	4.1

Example 117   Oxazole (C₃H₃NO)

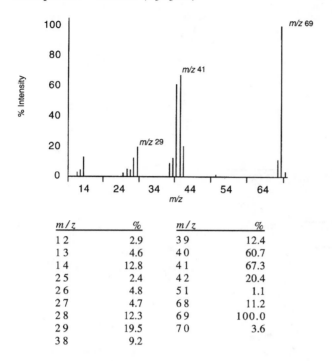

m/z	%	m/z	%
1 2	2.9	3 9	12.4
1 3	4.6	4 0	60.7
1 4	12.8	4 1	67.3
2 5	2.4	4 2	20.4
2 6	4.8	5 1	1.1
2 7	4.7	6 8	11.2
2 8	12.3	6 9	100.0
2 9	19.5	7 0	3.6
3 8	9.2		

Example 118   Pyridine (C₅H₅N)

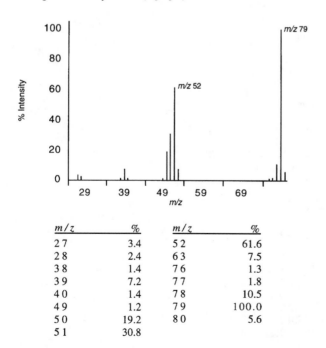

m/z	%	m/z	%
2 7	3.4	5 2	61.6
2 8	2.4	6 3	7.5
3 8	1.4	7 6	1.3
3 9	7.2	7 7	1.8
4 0	1.4	7 8	10.5
4 9	1.2	7 9	100.0
5 0	19.2	8 0	5.6
5 1	30.8		

Example 119   Thiophene (C$_4$H$_4$S)

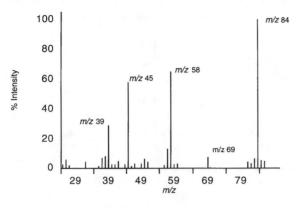

m/z	%	m/z	%	m/z	%
25	2.1	42	4.4	58	64.9
26	5.6	44	2.1	59	2.4
27	1.6	45	57.8	60	2.8
32	3.7	46	1.4	69	7.2
36	1.4	47	2.6	81	3.8
37	6.6	49	3.0	82	2.7
38	7.7	50	6.0	83	6.3
39	28.6	51	3.8	84	100.0
40	2.4	56	1.6	85	5.1
41	2.1	57	12.9	86	4.4

Example 120   Thiazole (C$_3$H$_2$NS)

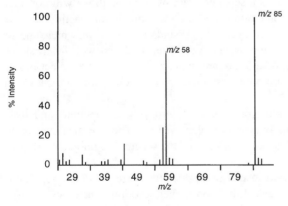

m/z	%	m/z	%
25	3.5	51	2.7
27	7.7	52	1.6
28	3.2	57	25.2
32	6.7	58	76.7
33	1.5	59	4.5
38	2.5	60	3.7
39	2.0	83	1.0
40	3.2	85	100.0
44	3.1	86	4.4
45	13.8	87	4.2

Example 121   Quinoline (C$_9$H$_7$N)

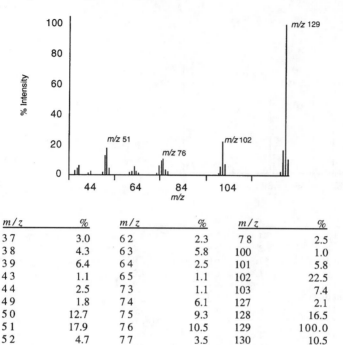

m/z	%	m/z	%	m/z	%
37	3.0	62	2.3	78	2.5
38	4.3	63	5.8	100	1.0
39	6.4	64	2.5	101	5.8
43	1.1	65	1.1	102	22.5
44	2.5	73	1.1	103	7.4
49	1.8	74	6.1	127	2.1
50	12.7	75	9.3	128	16.5
51	17.9	76	10.5	129	100.0
52	4.7	77	3.5	130	10.5
61	1.8				

Rather than treat each heterocycle as a separate class of compound, we will consider these compounds as a single group and summarize the most important reactions for the compounds and their derivatives. Heterocyclic compounds are those in which one or more =C—H groups in an aromatic ring system have been replaced with heteroatoms such as N, S or O. Heterocyclic compounds are aromatic, with the unbonded electrons of the heteroatom participating in the aromatic system. Typical heterocyclics are shown in Figure 12.1.

As we have seen in other aromatic compounds, heterocyclic compounds and their substituted analogs produce reasonably intense molecular ions. The five-membered ring heterocyclics exhibit similar ring cleavage patterns, and the scheme for thiophene (Figure 12.2) is typical.

Methyl-substituted heterocycles will lose H to produce the corresponding 'tropylium' ions, which are frequently the base peak. Ethyl and higher substituted heterocycles will undergo β-cleavage, losing R˙ and producing a tropylium ion. If there are γ-hydrogen atoms in the alkyl chain, the McLafferty rearrangement will occur. Examples of these reactions are shown in Figure 12.3.

The ring systems fragment by loss of small neutral molecules such as HCN, CH=NH, HC=O, HCS, CO and C$_2$H$_2$.

For substituted heterocycles ArX (where Ar represents the parent heterocyclic compound and X is a hydrogen or a non-alkyl functional group) varying amounts of Ar$^+$ will be observed. For heterocycles with the general formula ArCOX the corresponding ArCO$^+$ ions will produce intense lines. For heterocyclic esters of the formula ArCO$_2$X, there will be intense lines at ArCO$_2$H$^+$ if X contains a γ-hydrogen (McLafferty rearrangement).

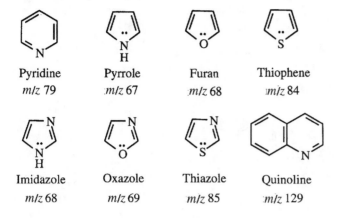

Pyridine	Pyrrole	Furan	Thiophene
m/z 79	*m/z* 67	*m/z* 68	*m/z* 84

Imidazole	Oxazole	Thiazole	Quinoline
m/z 68	*m/z* 69	*m/z* 85	*m/z* 129

Figure 12.1

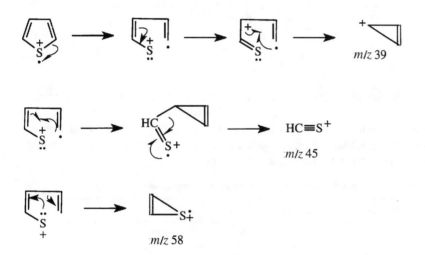

Figure 12.2

The spectrum of pyrrole (Example 114) exhibits the molecular ion at *m/z* 67, a line representing (M − HCN) at *m/z* 40 and a line at *m/z* 41 representing (M − CH = NH). Furan (Example 115) exhibits the molecular ion at *m/z* 68, an intense line at *m/z* 39 (M − CHO) and a relatively weak line at *m/z* 40 (M − C_2H_2). Imidazole (Example 116) exhibits the molecular ion at *m/z* 68 and a reasonably intense line at *m/z* 41 (M − HCN).

Oxazole (Example 117) exhibits the molecular ion at *m/z* 69. Loss of CHO˙ produces the ion at *m/z* 40, while *m/z* 41 and 42 correspond to loss of CO and HCN from the molecular ion. Pyridine (Example 118) exhibits loss of HCN from the molecular ion (*m/z* 79) as the dominant fragmentation. Thiophene (Example 119) exhibits two fragmentation lines of nearly equal intensity. The line at *m/z* 58 corresponds to loss of acetylene (C_2H_2) from the molecular ion, whereas the line at *m/z* 45 corresponds to SCH^+. The loss of SCH as a radical produces the line at *m/z* 39.

The loss of HCN followed by hydrogen loss produces the pair of lines at *m/z* 57 and 58 in thiazole (Example 120). Finally, in the spectrum of quinoline (Example 121), loss of

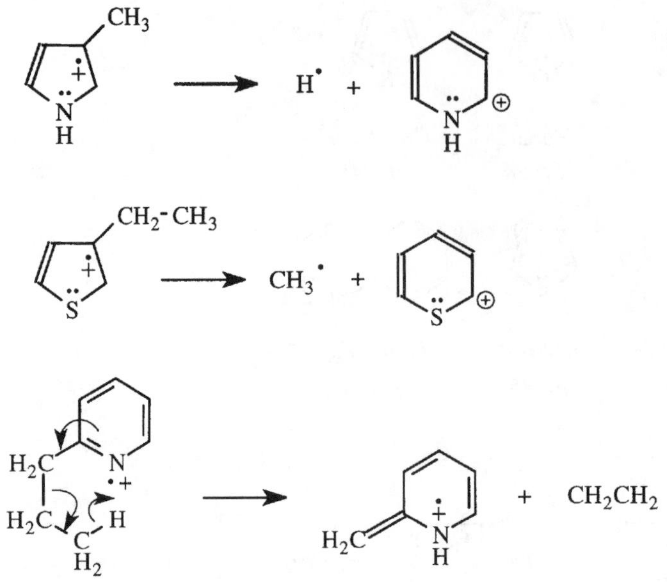

Figure 12.3

HCN produces the line at m/z 102, and loss of acetylene (C_2H_2) is responsible for the line at m/z 103.

The following table summarizes the typical m/z values encountered for the heterocyclics. Of course, for substituted heterocyclics these nominal m/z values will be increased depending on the nature of the substituent.

Table 5 Common fragments seen in heterocyclic compounds

Compound type	Neutrals lost[a]	m/z for		
		Ar^+	'Tropylium'	McLafferty
Pyridines	HCN	78	92	93
Quinolines	HCN	128	142	143
	C_2H_2			
Pyrroles	HCN	66	80	b
	CH=NH			
Furans	C_2H_2	67	81	b
	CHO			
Thiophene	C_2H_2	83	97	b
	SCH			
Imidazoles	HCN	67	81	b
Oxazoles	CHO	68	82	96
	HCN			
	CO			
Thiazoles	HCN	84	112^c	99
	HCN + H			

Notes:
[a] These are the most common losses, either as radicals or neutral molecules.
[b] β-Cleavage forming the tropylium ion predominates.
[c] Methylthiazoles do not form a tropylium ion. This peak is for the dialkylthiazoles.

13 Unknowns

Unknown No. 1

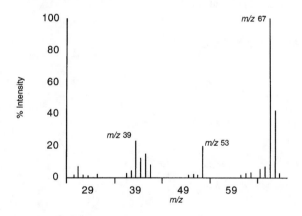

m/z	% Intensity	m/z	% Intensity
26	1.6	51	2.2
27	7.0	52	1.4
28	1.6	53	19.8
29	1.3	61	1.4
31	2.2	62	2.5
37	2.6	63	3.0
38	4.4	65	5.5
39	23.0	66	6.7
40	15.0	67	100.0
41	15.0	68	42.0
42	7.8	69	2.7
50	1.7		

134

Unknown No. 2

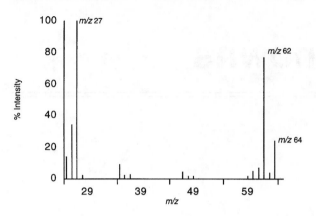

m/z	% Intensity	m/z	% Intensity
25	13.8	48	1.8
26	34.0	49	1.5
27	100.0	50	0.6
28	2.3	59	1.4
30	0.1	60	4.6
31	0.2	61	6.9
35	8.8	62	76.6
36	2.0	63	3.8
37	2.8	64	24.2
38	0.6	65	0.6
47	4.2		

Unknown No. 3

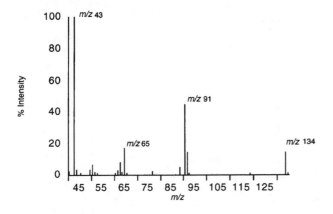

m/z	% Intensity	m/z	% Intensity
4 1	2.3	7 5	0.4
4 3	100.0	7 6	0.6
4 4	3.0	7 7	2.0
4 6	1.0	7 8	0.7
5 0	3.4	7 9	0.4
5 1	6.2	8 6	0.4
5 2	1.5	8 7	0.6
5 3	9.0	8 9	5.0
5 5	0.3	9 1	44.9
5 8	0.4	9 2	14.3
6 1	0.9	93	1.1
6 2	2.5	103	0.8
6 3	12.0	105	0.4
6 4	1.7	119	0.9
6 5	20.0	134	14.5
7 4	0.4	135	1.1

Unknown No. 4

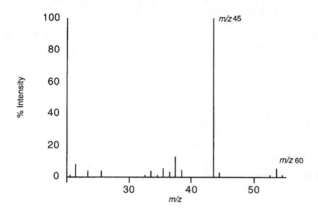

m/z	% Intensity	m/z	% Intensity
26	1.0	41	5.5
27	8.0	42	3.0
29	4.0	43	13.0
31	4.0	44	4.5
36	0.5	45	100.0
37	0.5	46	2.5
38	1.0	58	1.0
39	3.5	59	5.5
40	1.0	60	1.0

Unknown No. 5

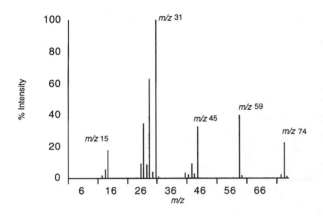

m/z	% Intensity	m/z	% Intensity
2	0.6	41	3.2
1 2	0.5	42	2.1
1 3	1.4	43	9.0
1 4	5.5	44	2.6
1 5	17.4	45	32.6
1 6	0.8	46	0.8
2 5	0.7	47	0.2
2 6	8.8	57	0.4
2 7	34.6	58	0.3
2 8	8.5	59	39.6
2 9	62.8	60	1.4
3 0	3.6	72	0.3
3 1	100.0	73	2.1
3 2	1.2	74	22.5
3 9	0.3	75	1.1
4 0	0.2		

Unknown No. 6

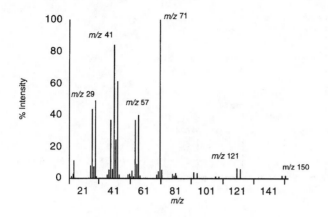

m/z	% Int	m/z	% Int	m/z	% Int
1 2	0.8	5 4	1.2	9 3	3.7
1 3	1.3	5 5	36.6	9 4	0.4
1 4	2.8	5 6	9.1	9 5	3.4
1 5	11.2	5 7	39.9	9 6	0.2
1 6	0.4	5 8	1.8	105	0.4
2 5	0.6	5 9	0.3	106	0.2
2 6	7.7	6 1	0.3	107	1.2
2 7	43.5	6 2	0.3	108	0.2
2 8	7.5	6 3	0.4	109	1.0
2 9	48.9	6 5	0.2	110	0.2
3 0	0.9	6 7	0.4	119	0.3
3 7	2.3	6 9	2.2	120	0.4
3 8	5.2	7 0	4.1	121	6.3
3 9	36.9	7 1	100.0	122	0.5
4 0	5.8	7 2	5.6	123	6.1
4 1	84.0	7 3	0.2	124	0.3
4 2	24.6	7 9	2.8	135	0.4
4 3	61.2	8 0	1.7	137	0.4
4 4	2.1	8 1	3.0	150	1.4
4 9	0.7	8 2	1.7	151	0.2
5 0	2.3	8 3	0.7	152	1.4
5 1	2.8	8 5	0.1	153	0.2
5 2	1.0	9 1	0.3		
5 3	4.7	9 2	0.3		

Unknown No. 7

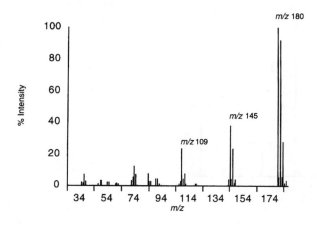

m/z	% Int	m/z	% Int	m/z	% Int
25	0.2	75	6.8	120	1.0
26	0.5	83	0.3	121	0.7
27	0.1	84	7.4	122	0.4
35	2.1	85	2.7	131	0.4
36	1.6	86	2.6	133	0.3
37	6.8	87	0.8	143	0.8
38	2.5	90	4.1	144	3.6
39	0.7	91	4.2	145	38.0
47	1.3	92	1.3	146	4.5
48	0.5	94	0.3	147	23.5
49	3.4	95	0.2	148	1.8
50	3.3	96	0.6	149	3.8
51	0.7	97	0.7	150	0.2
54	1.9	98	0.2	154	0.2
55	2.0	99	0.2	156	0.1
56	0.4	106	0.3	179	0.1
60	0.9	107	1.0	180	100.0
61	1.8	108	2.9	181	5.1
62	1.0	109	23.2	182	91.8
63	0.6	110	4.5	183	5.3
71	0.4	111	7.3	184	27.8
72	3.2	112	1.1	185	1.6
73	5.4	118	0.6	186	3.2
74	12.0	119	1.1	187	0.1

Unknown No. 8

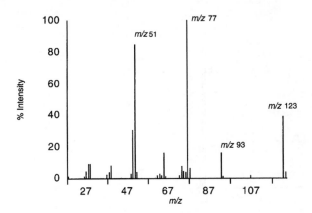

m/z	% Intensity	m/z	% Intensity
18	1.0	63	2.5
26	1.1	64	1.6
27	4.5	65	16.0
28	9.2	66	1.3
30	9.3	73	1.6
37	2.1	74	7.7
38	3.5	75	4.3
39	7.8	76	3.6
44	0.7	77	100.0
49	2.6	78	6.5
50	30.6	93	16.1
51	84.7	107	1.5
52	4.0	123	38.9
61	0.8	124	3.8
62	1.4		

Unknown No. 9

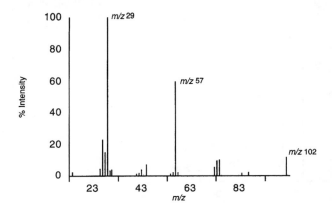

m/z	% Intensity	m/z	% Intensity
14	0.6	46	0.2
15	2.2	53	0.1
19	0.5	55	1.0
25	0.2	56	2.1
26	4.4	57	59.7
27	23.1	58	2.0
28	14.7	59	0.4
29	100.0	73	5.2
30	3.3	74	9.5
31	3.5	75	9.9
39	0.2	76	0.3
41	0.9	83	0.3
42	1.6	84	1.4
43	3.8	87	1.9
44	0.4	102	11.7
45	6.9	103	0.6

Unknown No. 10

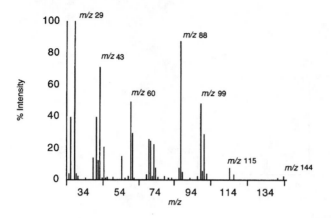

m/z	% Int	m/z	% Int	m/z	% Int
25	0.5	52	0.1	79	2.0
26	3.8	53	0.8	81	1.1
27	39.4	54	0.4	82	0.1
28	0.2	55	14.8	83	0.9
29	100.0	56	0.1	87	7.7
30	3.8	57	0.9	88	87.2
31	2.3	58	0.1	89	5.0
32	0.1	59	2.1	93	1.1
35	1.1	60	48.7	97	2.1
36	0.8	61	29.4	98	0.8
38	0.1	62	1.0	99	48.1
39	13.9	65	0.1	100	5.4
40	0.2	67	0.1	101	28.5
41	39.4	68	0.1	102	3.9
42	12.5	69	3.3	105	0.1
43	70.8	70	25.8	115	7.7
44	0.9	71	24.7	117	3.4
45	21.0	72	2.0	125	0.7
46	0.9	73	22.1	141	0.9
47	1.5	74	7.7	144	2.0
50	1.6	75	1.8	145	0.2
51	0.2				

Unknown No. 11

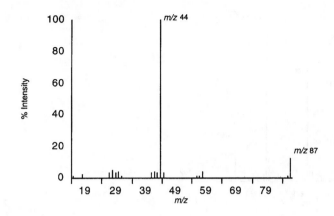

m/z	% Intensity	m/z	% Intensity
15	1.0	56	1.0
17	0.1	57	1.0
18	2.0	58	4.0
27	3.0	59	0.2
28	5.0	60	0.1
29	3.0	69	0.1
30	4.0	70	0.6
31	0.9	71	0.2
41	3.0	72	0.3
42	4.0	73	0.1
43	3.0	84	0.3
44	100.0	86	1.0
45	3.0	87	12.0
55	8.0		

Unknown No. 12

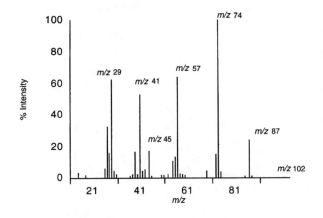

m/z	% Int	m/z	% Int	m/z	% Int
13	0.1	39	16.4	59	1.9
14	0.6	40	1.9	60	1.5
15	3.2	41	52.9	61	0.2
16	0.2	42	4.2	69	4.1
17	0.5	43	5.5	73	14.8
18	1.8	44	0.6	74	100.0
19	0.4	45	16.8	75	3.6
25	0.4	46	0.9	76	0.5
26	5.8	50	1.4	83	0.2
27	32.4	51	1.5	84	0.3
28	16.1	52	0.5	85	1.0
29	62.1	53	2.4	86	0.1
30	4.2	54	0.8	87	23.8
31	2.0	55	10.7	88	1.1
32	0.1	56	13.3	101	0.3
37	1.0	57	63.9	102	0.6
38	1.9	58	2.9	103	0.1

Unknown No. 13

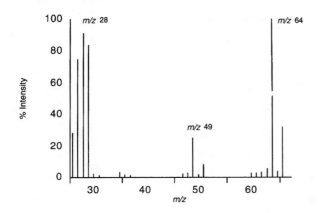

m/z	% Intensity	m/z	% Intensity
26	28.0	47	2.3
27	74.7	48	2.9
28	90.9	49	25.0
29	83.6	50	1.4
30	1.9	51	7.8
31	1.3	52	0.1
32	0.4	59	0.6
35	3.0	60	2.6
36	1.7	61	2.8
37	1.2	62	3.0
38	0.6	63	5.5
39	0.1	64	100.0
41	0.3	65	3.7
43	0.1	66	31.7
44	0.1	67	0.7

Unknown No. 14

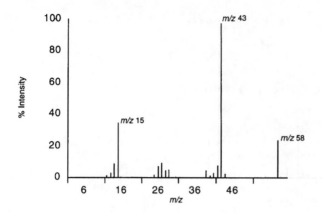

m/z	% Intensity	m/z	% Intensity
2	0.8	32	0.4
12	1.1	39	4.4
13	2.7	40	0.9
14	8.7	41	2.5
15	34.1	42	7.5
16	0.8	43	100.0
20	0.1	44	2.3
25	1.8	45	0.2
26	6.8	53	0.4
27	8.9	55	0.3
28	4.5	57	0.2
29	4.6	58	23.4
30	0.2	59	0.8
31	0.5		

Unknown No. 15

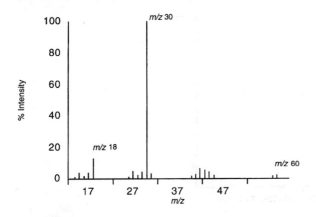

m/z	% Intensity	m/z	% Intensity
13	0.4	40	1.4
14	1.2	41	2.9
15	3.6	42	6.2
16	1.6	43	5.3
17	3.6	44	4.3
18	12.8	45	2.4
19	0.1	46	0.1
20	0.1	51	0.1
25	0.2	52	0.1
26	1.2	53	0.3
27	4.7	54	0.1
28	2.4	55	0.2
29	4.3	56	0.4
30	100.0	57	0.4
31	3.0	58	1.8
32	0.5	59	1.8
38	0.4	60	2.4
39	0.8	61	0.1

Unknown No. 16

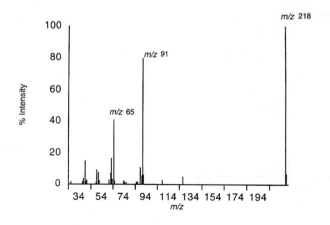

m/z	% Int	m/z	% Int	m/z	% Int
26	0.5	61	2.8	90	5.4
27	1.5	62	7.0	91	80.7
37	1.5	63	16.3	92	5.9
38	3.7	64	3.0	93	0.2
39	14.9	65	41.0	109	2.9
40	1.4	66	2.0	127	4.9
41	2.8	73	0.5	128	0.3
42	0.1	74	2.3	152	0.2
43	0.3	75	1.4	165	0.2
49	0.9	76	0.8	176	0.1
50	9.0	84	0.2	218	100.0
51	7.6	85	0.9	219	6.2
52	2.0	86	1.4	220	0.2
53	0.5	87	0.9		
60	0.2	89	10.8		

Unknown No. 17

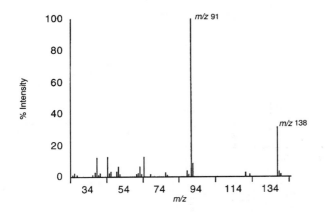

m/z	% Int	m/z	% Int	m/z	% Int
26	0.9	53	0.4	86	0.4
27	2.3	58	0.3	87	0.3
28	0.9	59	0.4	89	3.7
32	0.5	60	0.3	90	1.6
35	0.6	61	1.6	91	100.0
37	1.0	62	2.4	92	8.3
38	2.7	63	6.5	93	0.7
39	12.2	64	1.8	105	0.3
40	0.9	65	13.0	106	0.3
41	2.1	66	0.8	121	2.9
44	0.4	69	1.8	122	0.6
45	13.0	70	0.3	123	1.4
46	2.1	71	0.3	124	0.3
47	3.1	74	0.6	137	0.5
48	0.3	75	0.5	138	31.6
49	0.4	76	0.4	139	3.2
50	3.3	77	2.9	140	1.6
51	6.4	78	1.2	141	0.2
52	1.6	79	0.6		

Unknown No. 18

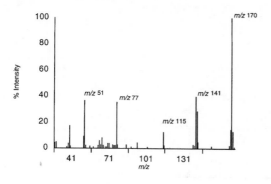

m/z	% Int	m/z	% Int	m/z	% Int
27	4.5	69	1.2	115	12.3
28	4.9	70	3.5	116	2.1
29	0.6	71	4.0	117	0.2
34	0.1	72	0.5	118	0.8
37	1.0	73	0.4	119	0.5
38	3.9	74	2.8	121	0.3
39	16.8	75	2.2	126	0.2
40	1.6	76	2.4	127	0.4
41	0.7	77	35.3	128	0.4
42	0.2	78	2.5	129	0.5
43	0.2	79	0.4	130	0.1
45	0.1	83	0.3	131	0.2
49	0.6	84	0.8	138	0.2
50	9.3	85	2.7	139	2.6
51	36.2	86	0.2	140	2.1
52	2.2	87	0.4	141	39.3
53	0.7	88	0.3	142	28.3
54	0.1	89	1.2	143	4.8
55	1.4	90	0.2	151	0.3
57	0.5	91	0.8	152	0.6
58	1.2	92	0.5	153	0.7
59	0.1	93	0.4	154	1.6
60	0.2	94	4.1	155	0.5
61	0.5	95	0.4	156	0.1
62	1.0	98	0.1	165	0.1
63	6.1	101	0.2	167	0.2
64	2.6	102	1.0	168	2.3
65	7.9	103	0.3	169	14.5
66	2.2	105	0.2	170	100.0
67	0.2	113	0.5	171	12.7
68	1.0	114	0.6	172	1.0

Unknown No. 19

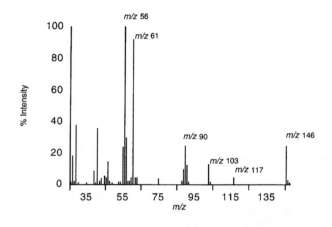

m/z	% Int	m/z	% Int	m/z	% Int
26	1.6	53	1.8	88	2.2
27	17.9	54	1.4	89	9.4
28	1.9	55	23.8	90	24.6
29	38.0	56	100.0	91	12.0
30	1.1	57	29.6	92	1.5
32	0.1	58	2.2	93	0.5
35	1.3	59	2.0	103	12.8
39	8.5	60	4.5	104	1.4
40	1.2	61	92.2	105	0.5
41	35.4	62	4.2	117	4.3
42	2.3	63	4.2	131	0.3
43	3.7	69	0.6	146	24.3
45	5.1	73	0.8	147	2.5
47	14.5	75	3.5	148	1.2
48	2.4	76	0.6		
49	1.3	87	0.5		

Unknown No. 20

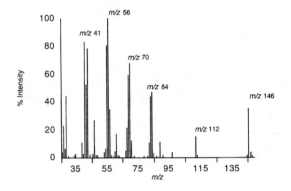

m/z	% Int	m/z	% Int	m/z	% Int
26	3.6	58	1.6	87	1.1
27	22.8	59	1.1	88	0.6
28	6.4	60	4.3	89	11.1
29	44.0	61	17.2	90	0.6
30	1.1	62	1.5	91	1.4
32	0.5	63	1.2	92	0.2
33	0.5	64	0.2	93	0.1
34	1.4	65	0.7	94	0.6
35	2.8	66	0.4	95	0.2
36	0.4	67	4.9	96	0.2
38	0.3	68	21.1	97	4.0
39	10.7	69	59.1	98	0.3
40	2.8	70	67.4	101	0.3
41	83.1	71	12.5	103	0.8
42	51.9	72	0.5	106	0.2
43	78.0	73	0.9	107	0.3
44	2.4	75	0.7	109	0.2
46	2.8	76	0.3	111	0.2
48	1.5	77	0.4	112	15.0
49	1.4	78	0.2	113	13.8
50	0.5	79	1.1	114	0.1
51	0.6	80	0.2	115	0.1
52	0.6	81	1.2	117	0.3
53	4.0	82	10.7	145	1.4
54	6.2	83	43.8	146	35.2
55	80.4	84	46.9	147	4.0
56	100.0	85	3.4	148	1.6
57	34.8	86	0.6	149	0.1

APPENDIX
Detailed Solutions to
Unknowns

Chapter 3

Problem 1

Inserting the proper error limits, we get the following results:

m/z	Abundance
44	100.0%
45	1.2 ± 0.2
46	0.4 ± 0.2

and from the $A+2$ line, we can have one or two oxygens (three oxygens is clearly impossible). The $A+1$ line indicates that we can have a maximum of one carbon. Since $CH_{14}O$ is nonsense, the only reasonable formula is CO_2. Please note: CNO is not possible because of the nitrogen rule. What about CH_4N_2? Calculating the saturation index we find

$$1 - 1/2(4) + 1/2(2) + 1 = 1.0$$

which seems unlikely for such a small molecule.

Problem 2

Inserting the proper error limits we see that:

m/z	Abundance
30	100.0%
31	2.3 ± 0.2
32	0.0 ± 0.2

and from the $A + 2$ line we find that we can have a maximum of one oxygen. From the $A + 1$ line, we have two carbons and therefore we cannot have any oxygens. The balance of the mass must be hydrogen, and we determine the formula to be C_2H_6 (saturation index $= 0.0$).

Problem 3

After inserting the appropriate error limits we find the following:

m/z	Abundance
84	100.0%
85	6.7 ± 0.7
86	0.0 ± 0.2

indicating a maximum of one oxygen atom. The $A + 1$ line covers the range 6.0–7.4%. This range is too high for five carbons (5.5%) and too low for seven carbons (7.7%), so it seems that there are only six carbons present. With six carbons, it is impossible to have any oxygen (and any nitrogens, by the nitrogen rule), so the balance of the mass is hydrogen. The formula is C_6H_{12} (saturation index $= 1.0$).

Problem 4

Once again, inserting the proper error limits gives us the following results:

m/z	Abundance
56	100.0%
57	4.7 ± 0.5
58	0.1 ± 0.2

and the $A + 2$ line indicates a maximum of one oxygen. The range of values for the $A + 1$ line is 4.2–5.2. This is too low for five carbons, but can be four carbons. Once again, we find that for four carbons neither oxygen nor nitrogen is possible, and the balance must therefore be hydrogen. The formula is C_4H_8 (saturation index $= 1.0$).

Problem 5

The first thing that I notice is that the 'molecular ion' has an odd mass. There are two possibilities: either this is not a molecular ion, or it contains an odd number of nitrogens. I am going to have to account for both possibilities. Inserting my error limits I find the following:

m/z	Abundance
91	100.0%
92	7.6 ± 0.8
93	0.3 ± 0.2

and from the $A + 2$ line I see that there can be a maximum of two oxygens and a minimum of zero oxygens. The values for the $A + 1$ line cover the range 6.8–8.4. This indicates that there are probably seven carbons present in this ion, although six carbons is remotely possible.

What do we know so far? Well, we know the fragment contains either six carbons (unlikely) or seven carbons, from zero to two oxygens and either one or three nitrogens (from the odd mass of the 'molecular' ion and the nitrogen rule). We should be able to develop reasonable formulae if we are consistent. Let us tabulate our results as we develop them.

Formula	C	N	O	Comments
1	6	0	2	Impossible
2	6	0	1	3 amu remaining
3	6	0	0	19 amu remaining
4	6	1	0	5 amu remaining
5	7	0	0	7 amu remaining

Of the four formulae developed (No. 1 being rejected because it is clearly impossible) we can reject No. 3 since it requires an impossible number of hydrogens (remember the $2n + 2$ rule). We can now calculate the saturation index for each of the remaining possible formulae.

Formula	R + DB	Comments
2. C_6H_3O	$6 - 1/2(3) + 1 = 5.5$	Indicates an even-electron ion and is very unsaturated
4. C_6H_5N	$6 - 1/2(5) + 1/2(1) + 1 = 5.0$	Consistent with an odd-electron ion, but very unsaturated
5. C_7H_7	$7 - 1/2(7) + 1 = 4.5$	Indicates an unsaturated even-electron ion.

Is there any way to select from the three possible formulae? We can probably make a reasonable attempt if we remember that benzene (C_6H_6) would have a saturation index of 4 (three double bonds and one ring). It is remotely possible that formula 2 is correct, but this is a highly unsaturated even-electron ion. It is also possible that formula 4 is correct, but this is also a highly unsaturated molecule. If formula 5 is the correct one, then the line at m/z 91 is not the molecular ion, but is instead a fragment ion.

Problem 6

m/z	Abundance
46	100.0%
47	2.9 ± 0.3
48	0.2 ± 0.2

The $A + 2$ line indicates a maximum of two oxygens, while the $A + 1$ line is consistent with two or three carbons. If nitrogen is present, we would have to have two nitrogen atoms, and it is very unlikely that we have both nitrogen and oxygen present in this molecule (28 from two nitrogens and 16 from one oxygen accounts for 44 mass units). Let us see what kinds of formulae we can propose.

Formula	C	N	O	Comments
1	2	0	1	6 amu remaining
2	3	0	0	10 amu remaining

The problem has been greatly simplified owing to the relatively low mass and the nitrogen rule. We can eliminate formula 2 because too many hydrogens are required (a C_3 hydrocarbon requires eight hydrogens), and therefore the correct formula is C_3H_6O (saturation index $= 1.0$).

Problem 7

m/z	Abundance
106	100.0%
107	7.6 ± 0.8
108	0.4 ± 0.2

The $A + 2$ line allows a maximum of three oxygens, while the $A + 1$ line strongly indicates seven carbons. With seven carbons, 84 amu are accounted for, leaving 22 amu to assign. Clearly, no nitrogen can be present and we must have one oxygen (seven carbons would require 16 hydrogens for maximum saturation). The only logical formula is C_7H_6O (saturation index $= 4.0$).

Problem 8

m/z	Abundance
93	100.0%
94	7.1 ± 0.7
95	0.2 ± 0.2

Once again, my attention is immediately focused on the odd mass value of the supposed molecular ion. If this is a molecular ion, then an odd number of nitrogens must be part of the molecular formula. The $A + 2$ line indicates a maximum of two oxygens, and the $A + 1$ line allows six or seven carbons. Taking these considerations into account, I can write the following formulae.

Formula	C	N	O	Comments
1	6	1	0	7 amu left
2	6	0	1	5 amu left
3	7	0	0	9 amu left

The calculation of the saturation index results in the following values.

Formula	R + DB	Comments
1. C_6H_7N	$6 - 1/2(7) + 1/2(1) + 1 = 4.0$	An odd-electron ion
2. C_6H_5O	$6 - 1/2(5) + 1 = 4.5$	An even-electron ion and therefore not the molecular ion
3. C_7H_9	$7 - 1/2(9) + 1 = 3.5$	An even-electron ion

Chapter 13

Unknown No. 1 solution

The general appearance of the spectrum indicates that this is a small, relatively stable molecule producing few fragments. The base peak is at m/z 67, but it is unlikely that this is the parent ion, because if it is the parent ion, then nitrogen is present and we should see many even mass fragments, which are missing from the spectrum.

This leads me to suspect that m/z 68 could be the parent ion. Normalizing m/z 68 and 69 and inserting the appropriate error limits gives me the following values:

m/z	% Intensity
68	100.0
69	6.4 ± 0.6

The range of values for m/z 69 is from 5.8 to 7.0%. The maximum number of carbons present would be six based upon the percentage natural abundance values, but six carbons would result in a mass of 72, which is clearly impossible. For the present, I will work from the assumption that there are only five carbons present in the molecule.

With five carbons present (for a mass of 60) the only other element that can be present is hydrogen, and I arrive at the formula C_5H_8. Calculating the number of rings and double bonds

$$\text{Rings} + \text{double bonds} = 5 - 1/2(8) + 1 = 2$$

I see that I can have one triple bond, two double bonds, one ring and one double bond, or two rings. It is unlikely that I will have two rings with only five carbons, so one of the other combinations is likely. I also see that, since the value for rings and double bonds is even, my proposed formula is an odd-electron ion and therefore could be the molecular ion.

It is now time to look at the losses from the parent ion, and I am going to consider those fragments at m/z 67, 53, 41 and 39, since they are the largest mass fragments of reasonable intensity in the spectrum.

m/z	Source	Compound indicated
67	Loss of H from parent	Various types such as aldehydes, acetals, aryl-CH_3, N-CH_3, $-CH_2CN$, and alkynes
53	Loss of 15 from parent	Various types such as acetals, methyl derivatives, t-butyl and i-propyl compounds, aryl-C_2H_5, $(CH_3)_2SiO$ derivatives
	Loss of 14 from m/z 67	Unlikely, not a logical loss
41	Loss of 27 from parent	Nitrogen-containing, such as aromatic amines or nitriles and nitrogen heterocycles
	Loss of 26 from m/z 67	Aromatic hydrocarbons

	Loss of 12 from m/z 53	Unlikely, not a logical loss
39	Loss of 29 from parent	Various types such as aromatic aldehydes, keto derivatives of cycloalkanes, naphthols, polyhydroxybenzenes, propionals, nitriles and ethyl derivatives
	Loss of 28 from m/z 67	Various types including phenols and naphthols, aldehydes, quinones, nitriles and ethyl esters
	Loss of 14 from m/z 53	Unlikely, not a logical loss
	Loss of 2 from m/z 41	Fused ring aromatic compounds

Unfortunately, this hasn't helped me very much since most of the compounds listed are impossible based on the molecular formula.

The next step is to try to identify the fragment ions.

m/z	Formula	Compound indicated
67	None listed	
53	None listed	
41	C_3H_5	Unsaturated hydrocarbon
39	C_3H_3	Hydrocarbons and especially aromatics

Conclusion

I am unable to uniquely identify this compound based upon the mass spectrum. I have calculated a reasonable formula for the compound based upon the isotopic ratios and have determined that it is an alkyne, a diene or a cyclic alkene.

Compound's actual identity: cyclopentene.

Unknown No. 2 solution

Examining the spectrum, I see that the compound has a low molecular weight and is relatively stable, producing a limited number of fragments. Almost immediately I notice the pair of lines at m/z 62 and 64 which appear to be due to chlorine. Normalizing these lines, I find that m/z 62 has a relative abundance of 100% and m/z 64 has a relative abundance of 31.5%, which matches the normal isotopic abundances for ^{35}Cl and ^{37}Cl quite nicely.

Having determined that a single chlorine atom is present, I am ready to calculate the number of carbons present, and I will use the pair of lines at m/z 64 and 65. I use these lines because it is possible that m/z 63 is contaminated by loss of hydrogen from m/z 64. Normalizing m/z 64 to 100% and inserting the appropriate error limits I get the following values.

m/z	% Intensity
64	100.0
65	2.5 ± 0.2

It is evident that there are two carbons present in the molecule. With a total mass of 61 (24 from the two carbons and 37 from ^{37}Cl) the formula for this compound must be C_2H_3Cl.

Calculating the value for rings and double bonds I get

$$2 - 1/2(4) + 1 = 1 \qquad \text{(note, halogens are counted as}$$
$$\text{hydrogen in this calculation)}$$

indicating that I have one double bond (the molecule is much too small to form a ring). Since the value for rings and double bonds is a whole number, this ion is an odd-electron ion and can be the parent compound.

There are no isomeric forms possible with this formula, so the compound must be chloroethene. The lines at m/z 26 and 27 represent loss of HCl and Cl respectively from the parent compound.

Compound's actual identity: chloroethene.

Unknown No. 3 solution

An inspection of the spectrum indicates a relatively stable molecule forming few fragments. With the exception of the even mass value at m/z 134, all fragment lines have odd mass values. This indicates (from the nitrogen rule) that there are either no nitrogens present or an even number of nitrogens. There is no chlorine, bromine or iodine present nor does there appear to be any sulfur or silicon.

My inspection shows me that there are four reasonably intense line fragments at m/z 43, 65, 91 and 134. The line fragments at m/z 65 and 91 are characteristic of the tropylium ion $(C_7H_7^+)$, indicating a benzene ring with an attached methyl group. The line fragment at m/z 43 is characteristic of either the propyl ion or CH_3CO^+, while the line at m/z 134 could be the molecular ion.

Re-normalizing the intensity values for m/z 134 and 135 and inserting the appropriate error limits gives me the following values:

m/z	% Intensity
134	100.0
135	7.5 ± 0.8

Indicating either six or seven carbons. This is inconsistent with the idea that m/z 43 and 91 result from simple cleavage of the molecular ion at m/z 134. If there are only seven carbons, then 50 mass units must be made up from the remaining elements (H, N, O, F and P).

Assuming for the moment that the fragments at m/z 43 and 91 are the result of simple cleavage of the molecular ion, then I can determine that there are either nine or ten carbons in the molecule. This lets me calculate two reasonable formulae for the molecular ion:

$$C_9H_7F \qquad R + DB = 9 - 1/2(8) + 1 = 6$$
$$C_9H_{10}O \qquad R + DB = 9 - 1/2(10) + 1 = 5$$

Since these are both whole numbers, either would be a legitimate formula for the molecular ion.

We can however, make a reasonable choice between the two formulae by looking once again at the fragmentation. If m/z 91 is in fact the tropylium ion ($C_7H_7^+$), then to form this ion from C_9H_7F would require us to lose C_2F, which is a very unlikely neutral. On the other hand, loss of C_2H_3O (or CH_3CO) from $C_9H_{10}O$ would be very reasonable.

Finally, we can combine our two fragment ions to identify our spectrum as being that of benzyl methyl ketone.

Compound's actual identity: benzyl methyl ketone.

Unknown No. 4 solution

An inspection of the spectrum indicates that we have a low molecular weight compound. The base peak in the spectrum is at m/z 45. Although there is a reasonably intense peak at m/z 59, it is unlikely that this is the parent: if m/z 59 is the parent then there is an odd number of nitrogens present and our spectrum should be dominated by even mass fragments. It would be more reasonable to suspect that m/z 60 is the parent ion.

There are no isotope patterns consistent with any $A+2$ element (other than oxygen). If nitrogen is present, there must be an even number of nitrogen atoms.

Since we do not have an isotope line at m/z 61, we cannot calculate the molecular formula in the traditional fashion. We should be able to get a fragment formula for m/z 45. Normalizing the data and inserting the appropriate error limits we get the following values:

m/z	% Intensity
45	100.0
46	2.5 ± 0.2

which is consistent with a fragment containing two carbons. The remaining mass (21 amu) must come from the remaining elements. This eliminates phosphorus from our list. The only remaining possibilities are fluorine, oxygen and nitrogen, and we can write the following fragment formulae:

$$C_2H_5O \qquad R+DB = 2 - 1/2(5) + 1 = 0.5$$
$$C_2H_7N \qquad R+DB = 2 - 1/2(7) + 1/2(1) + 1 = 0.0$$
$$C_2H_2F \qquad R+DB = 2 - 1/2(3) + 1 = 1.5$$

The whole number value for C_2H_7N indicates an odd-electron ion! This would be reasonable if m/z 45 were the molecular ion, but makes little sense in this case, since we have identified this peak as probably being a fragment ion and we can reject this formula.

The value for C_2H_2F indicates one double bond, which does not seem particularly likely since the fragment is so small. Another factor arguing against fluorine being present is a lack of peaks corresponding to loss of F and HF (19 and 20 respectively). Clearly the most reasonable formula for the fragment is C_2H_5O.

Consulting our ion composition table (Table 3), we find that our formula is reasonable.

If m/z 60 is the parent, then the most likely neutral lost in forming m/z 45 would be CH_3. This gives us a molecular formula of C_3H_8O, which has a $R+DB$ value of 0.0. The compound must be either an alcohol or an ether and we have the following possibilities:

1. Propanol
2. 2-Propanol
3. Methyl ethyl ether

I would be more inclined to suspect that the unknown is an alcohol rather than an ether. Even small ethers, such as dimethyl ether (Example 47) and methyl propyl ether (Example 48), have reasonably intense parent ion peaks. The relatively weak parent ion peak seen in this spectrum is more consistent with an alcohol than with an ether. Distinguishing between the two alcohols may be very difficult based solely upon the mass spectrum.

Compound's actual identity: 2-propanol.

Unknown No. 5 solution

Looking at the spectrum, I see a relatively small molecule producing many fragment peaks. This leads me to the conclusion that it is probably not an aromatic or other relatively stable molecule. The odd mass fragments indicate that there are either no nitrogens or an even number of nitrogens. The grouping of lines at about 14 amu increments indicate that the molecule may be a linear alkane, or a slightly branched alkane.

There is no chlorine, bromine or iodine present and there does not appear to be any silicon or sulfur. The peak at m/z 74 appears to be the parent ion. Re-normalizing and inserting the error limits for the line pair at m/z 74 and 75 gives me the following values:

m/z	% Intensity
74	100.0
75	4.7 ± 0.5

indicating the presence of four carbons.

The remaining mass (26 amu) must contain at least one non-hydrogen element, since C_4H_{26} is clearly nonsense. Phosphorus is eliminated from consideration owing to its mass of 31. I can also eliminate nitrogen from consideration since two nitrogen atoms give a total of 28 amu.

Two formulae can be written:

$$C_4H_7F \qquad R + DB = 4 - 1/2(8) + 1 = 1.0$$
$$C_4H_{10}O \qquad R + DB = 4 - 1/2(10) + 1 = 0.0$$

The fragment at m/z 59 corresponds to loss of CH_3 as a neutral from the parent. The fragment at m/z 45 corresponds to loss of CH_3CH_2. The peak at m/z 31 corresponds to $CH_2 = OH^+$, which is characteristic of methoxy compounds and primary alcohols. These fragments indicate that the molecule is not C_4H_7F.

For the formula $C_4H_{10}O$, we can have either an alcohol or an ether. With a relatively intense molecular ion, we can probably eliminate alcohols from consideration (see Examples 40 and 41). For the ethers, we have three possibilities:

1. Diethyl ether
2. Methyl propyl ether
3. Methyl isopropyl ether

Only diethyl ether could lose CH_3CH_2 without considerable rearrangement. I would identify this compound as diethyl ether.

Compound's actual identity: diethyl ether

Unknown No. 6 solution

Looking at the spectrum, I see many fragments occurring at intervals of 14 amu. This leads me to believe that this compound is probably some sort of alkane as opposed to a more stable aromatic compound. The unusually intense peak at m/z 71 compared with m/z 57 indicates that we have a branched alkane. What is very strange is that the intensity of the fragments above m/z 71 is very low; there is no smooth drop off with increasing mass.

At the high mass end of the spectrum, I notice a pair of lines of almost equal intensity at m/z 121 and 123 and a second pair of lines of almost equal intensity at m/z 150 and 152. The only common $A+2$ element with nearly equal abundance for both of its isotopic forms is bromine. There does not appear to be any other $A+2$ element (which would complicate the isotope pattern) so I am sure that there is no chlorine, sulfur or silicon.

I am unable to get a reasonable value for the number of carbons based on the isotope ratio of 153/152 (which works out to be about 14%). I notice that m/z 71 (corresponding to loss of ^{79}Br from m/z 150) is the base peak, and so I try to calculate the number of carbons present in this fragment. Re-normalizing the data and inserting the proper error limits I get the following values:

m/z	% Intensity
71	100.0
72	5.6 ± 0.6

which indicates the presence of five carbons in this fragment (and therefore in the parent molecule). These five carbons account for 60 amu, and the rest of the mass must be hydrogen.

I calculate the molecular formula as $C_5H_{11}Br$ and find that the value for rings and double bonds is 0.0. This is consistent with the molecular ion and indicates a saturated hydrocarbon. The fingerprint region indicates a single branch, so I conclude that this spectrum is of a monobrominated monomethylbutane. I am not able to determine the specific isomer.

Compound's actual identity: 1-bromo-2-methylbutane.

Unknown No. 7 solution

Examining the spectrum, I see few peaks in the low mass end, indicating a relatively stable molecule, perhaps an aromatic compound. At the high mass end, I see four lines at m/z 180, 182, 184 and 186. This indicates that there are multiple $A+2$ elements present in this compound, and I suspect that either bromine, chlorine or some combination of bromine and chlorine is present. This suspicion is further confirmed by the pattern of lines occurring at *m/z* 145, 147 and 149 and at *m/z* 109 and 111.

Using the rule that the number of A + 2 elements equals the number of isotope lines minus one, I suspect three A + 2 elements. I will now calculate the expected isotope patterns and compare them with my spectrum.

If all three of the A + 2 elements were chlorine, I would expect to see the following ratios:

m/z	180	182	184	186
Intensity	100.0	100.0	33.3	3.7

and these match sufficiently closely with the ratios in the spectrum for me to conclude that there are three chlorine atoms present.

Subtracting the mass of three ^{35}Cl isotopes from 180 leaves me 75 mass units for which to account.

Based upon the regular nature of the isotope pattern produced by the three chlorine atoms, and by the residual mass, I can conclude that there is no bromine, iodine, sulfur or silicon in this compound.

I am now ready to calculate the number of carbon atoms present. For this calculation, I will use the line pair at *m/z* 186 and 187. I choose this line pair because I know that halogenated materials can readily lose hydrogen, and this would cause the lower mass line pairs to be contaminated. Re-normalizing and inserting the error limits gives me the following results:

m/z	% Intensity
186	100.0
187	4.4 ± 0.4

which indicates that there are four carbons present. I am somewhat suspicious of this value, as it seems somewhat low, and I remember that sometimes the abundances may be slightly off due to the relatively low intensities of the lines with which I am working. I will use the value of four carbons as a reasonable lower limit and calculate various formulae:

C	Cl	N	O	F	P	H	
4	3	–	–	–	–	27	Impossible[1]
4	3	–	–	–	1	–	Impossible[2]
4	3	–	–	1	–	8	Impossible[1]
4	3	–	1	–	–	11	Impossible[1]
4	3	2*	–	–	–	–	Impossible[2]
5	3	–	–	–	–	15	Impossible[1]
5	3	–	–	1	–	–	Impossible[2]
5	3	–	1	–	–	–	Impossible[2]
6	3	–	–	–	–	3	

[1] Violates $2n + 2$ rule for saturated compounds.
[2] Mass exceeds 180.

For nitrogen, we must have an even number of nitrogen atoms, since we have an even mass. Logically, if four carbons, three chlorine and two nitrogens are impossible (the mass is too large) then there is little sense in testing five carbons, three chlorines and two nitrogens. Similar reasoning applies to the phosphorus case.

This leaves me with one molecular formula which can be the molecular ion. The compound contains a total of four rings and double bonds. A substituted benzene is the only realistic candidate.

I identify this spectrum as being of trichlorobenzene. Once again, I am unable to identify the specific isomer.

Compound's actual identity: 1,2,3-trichlorobenzene.

Unknown No. 8 solution

The spectrum indicates that this compound is relatively stable. The base peak at m/z 77, in conjunction with the peak at m/z 51, is characteristic for a benzene ring. At the upper mass end I see an intense peak at m/z 123, while the highest mass peak is at m/z 124. Immediately, I have a serious problem to resolve: is the peak at m/z 124 the molecular ion or is it a satellite of the molecular ion peak at m/z 123? If the peak at m/z 124 is the molecular ion, and if there are at least six carbons (from the benzene ring), then I would expect to see about 0.2% intensity at m/z 125. I do not see any intensity at m/z 125. On the other hand, if the peak at m/z 123 is the molecular ion peak, then I have an odd number of nitrogens present in this compound. The peak at 124 is the isotope satellite of that at m/z 123.

Calculating the number of carbons present based on the ratio of 124/123 (assuming a single nitrogen is present) does not yield a sensible result, as the 124 peak is 9.7% of the 123 peak, indicating eight or nine carbons.

There seem to be two possible cases: either m/z 77 comes from the parent ion at m/z 123, or it comes from the parent ion at m/z 124. In the first case, the mass difference is 46, in the second case the mass difference is 47.

Consulting our table of neutral losses, we find that a loss of 46 is commonly associated with loss of water followed by loss of C_2H_4, or loss of NO_2 or loss of CH_2S. As there is no evidence for the presence of sulfur, we can discount the last. Loss of 47 is commonly associated with loss of C_2H_4F and CH_3S. Once again, we can discount the latter.

What about the peak at m/z 93? Consulting our table of neutral losses, we see that a loss of 30 (123 − 93) corresponds to aromatic nitro compounds or aromatic methyl ethers. A loss of 31 (124 − 93) corresponds to methoxy compounds or methyl esters.

The only consistent type of compound indicated by the neutral losses is a nitro compound. I would identify this material as nitrobenzene.

Compound's actual identity: nitrobenzene.

Unknown No. 9 solution

A general inspection of the spectrum leaves the impression that we are dealing with a hydrocarbon. If any nitrogen atoms are present, there must be an even number of them since the spectrum is dominated by odd mass fragments and m/z 102 is clearly a candidate for the molecular ion peak. No chlorine, bromine, iodine, sulfur or silicon is present.

Re-normalizing the data for m/z 102 and 103 and inserting the appropriate error limits gives me the following values:

m/z	% Intensity
102	100.0
103	5.1 ± 0.5

which indicates that five carbons are present. Clearly, there must be at least one heteroatom present, since a saturated five carbon compound would have a mass of 72, leaving 30 mass units for which I have to account. The remaining heteroatoms available are nitrogen, oxygen, fluorine and phosphorus, and I construct the following table.

C	N	O	F	P	H	
5	–	–	–	1	11	
5	–	–	1	–	23	Impossible, saturation
5	–	–	2	–	4	
5	–	1	–	–	26	Impossible, saturation
5	–	1	1	–	7	
5	–	2	–	–	10	
5	2	–	–	–	14	

Testing the formulae for the number of rings and double bonds gives me the following results:

$$C_5H_{11}P \quad R+DB = 5 - 1/2(11) + 1/2(1) + 1 = 1.0$$
$$C_5H_4F_2 \quad R+DB = 5 - 1/2(6) + 1 = 3.0$$
$$C_5H_7OF \quad R+DB = 5 - 1/2(8) + 1 = 2.0$$
$$C_5H_{10}O_2 \quad R+DB = 5 - 1/2(10) + 1 = 1.0$$
$$C_5H_{14}N_2 \quad R+DB = 5 - 1/2(14) + 1/2(2) + 1 = 0.0$$

All of these values are consistent with the molecular ion, so I will have to look at the fragmentations. Looking at the neutral losses I see the following:

Neutral lost	Compound indicated
$102 - 57 = 45$	C_2H_5O—carboxylic acids, ethyl esters, ethers
$102 - 29 = 73$	Nothing common
$57 - 29 = 28$	C_2H_4, N_2, CO—phenols, aldehydes, diaryl ethers, quinones, aliphatic nitriles, ethyl esters

From the neutral losses, it appears that the only consistent compound would be some type of ethyl ester. This is further confirmed by consulting the ion fragment table for m/z 29 and 57:

m/z	Formula
29	$C_2H_5^+$, CHO^+
57	$C_4H_9^+$, $C_2H_5CO^+$

An ethyl ester would leave a three carbon acid, so I identify this compound as ethyl propanoate.

> Compound's actual identity: ethyl propanoate (propanoic acid ethyl ester).

Unknown No. 10 solution

A general inspection of the spectrum gives me the impression that this is some sort of linear or branched hydrocarbon chain. Two features attract my attention: the even mass fragments at m/z 60 and 88 and the pair of lines at m/z 99 and 101. Even mass fragments are potential odd-electron ions, and at first glance the fragments at m/z 99 and 101 suggest the possibility that there is a chlorine in this molecule.

To test the possibility of chlorine, I calculate the ratio of 99/101 and find that this value is 1.7 : 1.0. I can dismiss all thoughts of chlorine, since the real ratio of ^{35}Cl : ^{37}Cl is about 3 : 1.

The next thing to do is to see if there is a possible molecular ion. That at m/z 144 is a likely candidate. The absence of A + 2 lines indicates that there is no sulfur or silicon. Of the elements commonly found in organic compounds, I have reduced the possibilities to nitrogen, oxygen, fluorine, phosphorus and possibly iodine.

My next task is to see if I can determine the number of carbon atoms present. Re-normalizing the intensities for the 144/145 pair and inserting the appropriate error limits gives me the following values:

m/z	% Intensity
144	100.0
145	10.0 ± 1.0

I shall have to be careful with these values because of the relatively low intensities of the lines. Taking the error limits into account, I see that I could have eight, nine or ten carbons. I can definitely eliminate iodine, since its mass of 127 would be completely inconsistent.

Now, normally, I would go through the process of listing systematically all of the possible molecular formulae, eliminating those that are too massive or exceed reasonable saturation levels, testing for the number of rings and double bonds, etc. But I think in this case I will see if those ions at m/z 60 and 88 can provide a short-cut solution. I remember that one common source of odd-electron fragments is the McLafferty rearrangement, and so I consult my list of common McLafferty rearrangement peaks (page 89). I find that a peak at m/z 88 is commonly associated with an ethyl ester. The peak at m/z 60 would be consistent with a second McLafferty rearrangement, eliminating ethene from m/z 88.

It is clear to me that this is an ethyl ester of some sort. This would explain the peak at m/z 99, since esters normally lose the alkoxy group (actually, C_2H_5O in this case). The peak at m/z 99 must have the general formula $R-CO^+$ (the acid portion of the ester). Subtracting 28 (the mass of CO) from 99 gives me 71 mass units for which to account. Based on the number of carbons determined by the isotopic abundance, the only way to account for this mass is C_5H_{11}.

Putting these parts back together, I see that my ester takes the form of $C_5H_{11}COOC_2H_5$. The regular nature of the spectrum suggests little or no branching. I would identify this spectrum as that of ethyl hexanoate, although some sort of slightly branched analog is possible.

Compound's actual identity: ethyl hexanoate (hexanoic acid ethyl ester).

Unknown No. 11 solution

The combination of an even mass base peak at m/z 44 and an odd mass molecular ion candidate at m/z 87 convinces me that this compound contains at least one nitrogen atom. With so little information with which to work (only two peaks of reasonable intensity and no isotopic data for the molecular ion), I am going to have to rely on my tables of neutral losses and ion compositions.

From Table 2, the most common neutrals associated with a loss of 43 are HNCO, CH_3CO and C_3H_7. From Table 3, the most common ions with a mass of 44 are $[CH_2CHO + H]^+$, $CH_3CHNH_2^+$, $NH_2C=O^+$ and $(CH_3)_2N^+$. At first glance this doesn't seem to help much and I am still left with the problem of deciding if this is an amine or an amide. A general review of amide and amine fragmentation (Chapter 10) tells me that amides tend to form $NH_2C=O^+$ (and higher series if the nitrogen is substituted), that the acylium ion (RCO^+) is generally weak and that the McLafferty rearrangement occurs if the acid chain is three carbons long (or longer). How does this information help me?

If the spectrum is of an amide, then it should have the formula C_4H_9ON. It must be a simple amide (no alkyl groups on the nitrogen). The carbon chain must be branched, since if it was linear the McLafferty rearrangement would produce a peak at m/z 59 in my spectrum. The only possible amide consistent with this would be $(CH_3)_2CHCONH_2$ (2-methylpropanamide). The normal α-cleavage mechanism would produce the base peak at m/z 44 by ejection of C_3H_7 as a neutral.

If the spectrum is of an amine, I would expect it to be a secondary or tertiary amine, since a primary amine would give a molecular ion of relatively low intensity and would produce a significant fragment at m/z 30 ($CH_2=NH_2^+$). Considering the normal α-cleavage mechanism, there can be only one substituent on the nitrogen and that must be a methyl group. The remaining mass (57 amu) would be an n-butyl, isobutyl or t-butyl group.

We have five candidate molecules that are consistent with the relatively simple spectrum:

$$CH_3-CH-\overset{\overset{\displaystyle O}{\|}}{C}-NH_2$$
$$\underset{\displaystyle CH_3}{|}$$

$$CH_3CH_2CH_2CH_2\underset{\underset{\displaystyle H}{|}}{N}-CH_3 \qquad CH_3\underset{\underset{\displaystyle CH_3}{|}}{C}HCH_2NHCH_3$$

$$CH_3CH_2\underset{\underset{\displaystyle CH_3}{|}}{C}H-NHCH_3 \qquad CH_3-\overset{\overset{\displaystyle CH_3}{|}}{\underset{\underset{\displaystyle CH_3}{|}}{C}}-NHCH_3$$

Compound's actual identity: *N*-methylbutanamine.

Unknown No. 12 solution

Examining the spectrum, I see many fragments indicating that this molecule is probably a linear or branched material as opposed to an aromatic compound. My attention is drawn to three definite features of this spectrum: the relatively low intensity 'parent' at m/z 102, the odd-electron peak at m/z 74, and the presence of a significant fragment at m/z 45.

With the base peak occurring at m/z 74, I think that this is some sort of McLafferty rearrangement fragment. Consulting my list of McLafferty fragments, I find that a methyl ester would be consistent with this fragment.

There does not appear to be any significant A + 2 element present, other than oxygen, although I must admit that the intensity of the 'parent' at m/z 102 is very low. Using the intensity values of m/z 102 and 103 to try to calculate a molecular formula is not helpful.

Finally, m/z 45 is interesting because this is a common fragment to see in the spectra of acids. If the material were an ordinary straight chain acid, the McLafferty rearrangement peak would occur at m/z 60. Can I think of an explanation for this discrepancy? Well, if the material were branched at the second carbon this might explain the discrepancy, since the mass of $CH(CH_3)=C(OH)_2^+$ would be the same as that of $CH_2=C(OH)OCH_3^+$. Also, the low intensity of the supposed parent at m/z 102 is more consistent with the material being an acid (acids are notorious for having low intensity parents) than it would be if the material were an ester (which generally have reasonably intense parents).

It seems, then, that this spectrum is of a branched five carbon acid and I would tentatively identify it as 2-methylbutanoic acid. In this case, I would definitely want some other verification of this compound's identity, since so much of the mass spectral interpretation is based on 'ifs'; 'if' m/z 102 is the parent, 'if' m/z 74 is an acid McLafferty rearrangement, 'if' there are no elements other than C, H and O.

Compound's actual identity: 2-methylbutanoic acid.

Unknown No. 13 solution

In looking at this spectrum my attention is immediately drawn to the pair of lines at m/z 64 and 66. The ratio of these lines indicates the presence of a single chlorine atom.

Further inspection of the isotopic data indicates that this is probably the only $A+2$ element (with the possible exception of oxygen) present in this molecule.

The pair of lines at m/z 49 and 51 indicates that the chlorine atom is present on this fragment. The loss of 15 is consistent with loss of a methyl from the parent. The lines at m/z 28 and 29 correspond to loss of HCl and Cl (both with the ^{35}Cl isotope).

From simple inspection I am prepared to identify this compound as chloroethane. The patterns produced by the logical losses are really the only evidence that I need for this identification. For those students requiring more rigorous proof, I would point out that the ratio of m/z 69/68 is 2.2, indicating a total of two carbons. These two carbons and one chlorine account for 59 amu, with the balance accounted for by five hydrogens. The only reasonable chemical formula is C_2H_5Cl.

Compound's actual identity: chloroethane.

Unknown No. 14 solution

Clearly, this is the spectrum of a relatively small molecule. From the isotopic abundances, the only possible $A+2$ element would be oxygen. Nitrogen (if present) would occur as pairs of nitrogen atoms. Fluorine and phosphorus could also be present, although if fluorine is present then this molecule is unusual: fluorinated compounds generally show very weak parent ions.

Re-normalizing the data for the m/z 58, 59 pair and inserting the proper error limits gives me the following results:

m/z	% Intensity
58	100.0
59	3.4 ± 0.3

which indicates the presence of three carbons and eliminates the possibility of any phosphorus in the compound ($36 + 31 = 67$). The nitrogen rule eliminates the possibility of a pair of nitrogens ($36 + 28 = 64$).

With such a small molecule, writing out all the possible molecular formulae is not an overwhelming task. Here are the possibilities:

C	O	F	H	
3	–	–	22	Impossible, saturation
3	–	1	3	
3	1	–	6	

It appears that I have a choice of two formulae: either C_3H_3F or C_3H_6O. Calculating the rings and double bonds gives me the following values:

$$C_3H_3F \qquad 3 - 1/2(4) + 1 = 2.0$$
$$C_3H_6O \qquad 3 - 1/2(6) + 1 = 1.0$$

I have my choice of either a fluorinated diene (cyclopropene would probably be much too unstable owing to ring strain to be considered), an aldehyde or a ketone, or perhaps cyclopropanol.

Looking at m/z 43, I would suspect loss of methyl from the parent as the most likely source. On this basis, I could reject the fluorinated diene and cyclopropanol. Calculating the formula of the fragment at m/z 43 I find that the re-normalized data give me:

m/z	% Intensity
43	100.0
44	2.3 ± 0.2

indicating two carbons present in this fragment. The most likely formula is C_2H_3O (I just can't bring myself to believe that it would be $C=CF^+$).

Finally, distinguishing between aldehyde and ketone is relatively straightforward. For a molecule of this size, I would expect to see a significant peak at m/z 29 if the compound were an aldehyde (due to CHO^+). This peak is virtually absent from the spectrum (4.6% intensity). The only three carbon ketone is 2-propanone (otherwise known as acetone).

Compound's actual identity: 2-propanone (acetone).

Unknown No. 15 solution

What do we have here? This is quite a simple looking spectrum. I see the base peak at m/z 30 and not much else; a small peak at m/z 18 (could this be water?) and a possible parent at m/z 60. If the parent is m/z 60, then the base peak is an important odd-electron ion (so is m/z 18). I have no McLafferty peak for m/z 30 on my list. I also see no indication of A + 2 elements (with the exception of oxygen).

What happens when I re-normalize the m/z 60 and 61 peaks? I get the following values:

m/z	% Intensity
60	100.0
61	4.2 ± 0.4

which indicates four carbons. This would leave 12 amu to account for, and hydrogen would be the only reasonable choice. But the formula produced by this is impossible!

What about using the data for the base peak? Re-normalizing the data I get the following values:

m/z	% Intensity
30	100.0
31	3.0 ± 0.3

which indicates three carbons, and this is completely impossible!

So, it appears that my simple spectrum isn't so simple after all. I guess I could try and list all possible chemical formulae, but I'm not even sure that m/z 60 is the parent.

Maybe the ion fragment formula table can help me. Looking up the masses, I see that m/z 18 is commonly H_2O^+ or NH_4^+, while m/z 30 is commonly NO^+ or $CH_2NH_2^+$. With three out of four common fragments indicating the presence of nitrogen, it seems reasonable to be looking for a nitrogen-containing compound. If m/z 60 is the parent, then the compound must contain at least two nitrogens (four nitrogens would of course be pushing the absolute limit).

If this is the case, that m/z 60 is the parent ion, what could the rest of the molecule be? One possibility is that I have a symmetrical molecule, with the unknown being N_2O_2 or $C_2H_8N_2$. I guess I had better look at the table of neutral losses and see what a loss of 30 could be. As I suspected, both NO and CH_2NH_2 are on this list.

There are two possibilities: either m/z 60 is the parent ion, or the parent ion is not present in the spectrum.

If the parent ion is not present in the spectrum, then I have taken this identification process as far as possible and the base peak is either NO^+ or $CH_2NH_2^+$.

If m/z 60 is the parent ion, then I have to be dealing with something like N_2O_2 or $C_2H_8N_2$. This is consistent with both the ion fragment composition and the neutral loss. It is also consistent with the idea that if this odd-electron fragment contains a nitrogen, then the parent molecule must contain two nitrogens, otherwise m/z 60 would not be the parent ion and we're back to the case that the parent isn't in the spectrum.

Is there anything else that I can do? Well, there is one possibility. I have two different formulae for the (supposed) parent ion. I could calculate the expected isotope ratio and compare this with the actual isotope ratio. This might help me further distinguish between the two possibilities.

For N_2O_2 I can tabulate the following values:

Element	% Int 'A'	% Int 'A + 1'	% Int 'A + 2'
N_2	100.0	0.74	0.0
O_2	100.0	0.08	0.4
Total	100.0	0.82	0.4

These values don't compare particularly well with the measured ratios for m/z 60 and 61.

For $C_2H_8N_2$ I can repeat the procedure and I get the following values:

Element	% Int 'A'	% Int 'A + 1'	% Int 'A + 2'
C_2	100.0	2.2	0.0
H_8	100.0	0.12	0.0
N_2	100.0	0.74	0.0
Total	100.0	3.1	0.0
Spectrum	100.0	4.2	0.0

Although the match is much better in this case, I am still not completely certain that I have identified the compound. So much for this being a simple spectrum!

Compound's actual identity: ethylenediamine ($H_2NCH_2CH_2NH_2$).

Unknown No. 16 solution

When I look at this spectrum my attention is immediately drawn to the pair of lines at m/z 65 and 91. This line pair is characteristic of the tropylium ion, indicating a methylbenzene. I can't help but notice the peak at m/z 218. Could this base peak also be the molecular ion peak? The third feature of interest in this spectrum is the area between the tropylium ion (m/z 91) and the base peak at m/z 218. What is so interesting about this region? It is almost completely empty! The only significant fragments in this region are at m/z 109 and m/z 127.

I do not see any indication of A + 2 elements (oxygen is the exception of course). Let's see what I can do about calculating the molecular formula. Re-normalizing the data for m/z 218, 219 and 220 gives me the following values:

m/z	% Intensity
218	100.0
219	6.2 ± 0.6
220	0.2 ± 0.2

This verifies that there may or may not be an oxygen present (which I knew already) and indicates that there are probably six carbons.

But wait a minute. The tropylium ion has a formula of C_7H_7. This is inconsistent with six carbons and either my identification of the tropylium ion is wrong or for some reason my molecular formula calculation is in error. I could use the isotopic ratios for m/z 91 and 92 and determine an ion fragment formula, or I could use my table of neutral losses and see what kind of neutral has been lost. Finding the neutral loss is the easiest, and I see that a loss of 127 corresponds to iodine.

The presence of iodine in the molecule helps to explain the problem with calculating the molecular formula: halogenated compounds often give inconsistent results of this kind. It also explains the presence of a peak at m/z 127 (I^+).

Now I am ready to put the pieces together. I have a tropylium ion (probably from a methylbenzene fragment) and an iodine atom. The simplest compound for this unknown would be one of the isomeric iodomethylbenzenes. Determining the exact isomer is problematical from the spectrum, but since there are only three possibilities, this task could be accomplished by other means.

Compound's actual identity: 1-iodo-3-methylbenzene.

Unknown No. 17 solution

Once again I have a reasonably simple spectrum, dominated by the base peak at m/z 91. That this peak is due to the tropylium ion is confirmed by the presence of a second peak at m/z 65. The reasonably strong peak at m/z 138 is a candidate for the molecular ion. There does not appear to be any significant chlorine or bromine present. The pattern of lines at m/z 138, 139, 140 and 141 is suggestive of some sort of A + 2 element.

Re-normalizing the data and inserting the proper error limits gives me:

m/z	% Intensity
138	100.0
139	10.1 ± 1.0
140	5.0 ± 0.5
141	0.6 ± 0.2

The A + 2 line (m/z 140) is consistent with the presence of a sulfur atom (4.4% isotopic abundance). Since sulfur has three common isotopes, I think I will correct the data for the presence of one sulfur atom and I get the following results:

m/z	% Intensity	− S
138	100.0	100.0
139	10.1 ± 1.0	9.3 ± 0.9
140	5.0 ± 0.5	0.6 ± 0.2
141	0.6 ± 0.2	0.6 ± 0.2

This still leaves me with some $A + 2$ intensity and some intensity at the $A + 3$ position (m/z 141). With eight or nine carbons (based on the 139 intensity), I would expect some intensity at the $A + 2$ position from the possibility of having more than one ^{13}C isotope in any given molecule. This would account for 0.34–0.44% intensity at the $A + 2$ position, and correcting the m/z 140 line for this possibility reduces the intensity to the point that the only other possible $A + 2$ element would be oxygen (which I initially assume can be present). Given that I can have multiple ^{13}C in this molecule, and the presence of sulfur (having three isotopic forms), the relatively minor intensity at the $A + 3$ position makes sense.

If I have nine carbon atoms and one sulfur atom, my mass is 140, which exceeds the mass of the molecular ion (138), so clearly I must have eight carbons. The only formula which makes sense is $C_8H_{10}S$. Calculating the number of rings and double bonds gives me

$$8 - 1/2(10) + 1 = 4$$

which indicates (among other possibilities) a benzene ring, and would be consistent with the tropylium ion observed.

My compound is either a thiol or a thioether (sulfide). If it was a thiol, I would expect to see (among other possibilities) the loss of H_2S $(M - 34)$. The absence of any significant peak at m/z 104 leads me to believe that this compound is probably a thioether (sulfide).

So, what do I know? I know that I have a sulfur atom with methylbenzene attached to one side and a methyl attached on the other side. Two structures occur to me that would account for this spectrum:

Benzyl methyl sulfide

Methylbenzene methyl sulfide

Compound's actual identity: Benzyl methyl sulfide.

Unknown No. 18 solution

This spectrum is of a relatively large molecule. There are relatively few peaks, suggesting a stable molecule (aromatic?) instead of an aliphatic molecule. The presence of peaks at m/z 77 and 51 indicates a benzene ring. The absence of peaks at m/z 91 and 65 indicates that the benzene ring does not contain a methyl group. The base peak is m/z 170, which also appears to be a reasonable candidate for the molecular ion.

There does not appear to be any significant $A+2$ element present, with the exception of oxygen. If m/z 170 is the molecular ion, then the high intensity would tend to indicate that the molecule is not halogenated (halogenated compounds normally show reduced intensity); however, this possibility cannot yet be completed eliminated.

Re-normalizing the data for the lines from m/z 170 to 172 and inserting appropriate error limits gives me the following values:

m/z	% Intensity
170	100.0
171	12.7 ± 1.3
172	1.0 ± 0.2

which indicates either 11 or 12 carbons. The presence of so many carbons would account for most of the intensity of the 172 peak: 11 carbons would account for 0.67% while 12 carbons would account for 0.8%. If the molecule contains 11 carbons, then I have 38 mass units for which to account; if it contains 12 carbons, then I have 26 mass units for which to account. In any case, I can eliminate iodine from consideration.

I can begin writing potential formulae for the molecular ion:

C	H	O	N	F	P	
11	7	–	–	–	1	
11	19	–	–	1	–	
11	–	–	–	2	–	Impossible, saturation
11	10	–	2	–	–	
11	3	1	–	1	–	Impossible, saturation
11	22	1	–	–	–	
11	6	2	–	–	–	
11	38	–	–	–	–	Impossible, saturation
12	7	–	–	1	–	
12	10	1	–	–	–	
12	26	–	–	–	–	

Based upon the possible formulae, I have eliminated phosphorus from consideration. For the remaining possible formulae, I can calculate the number of rings and double bonds:

$$C_{11}H_7P \quad\quad R + DB = 11 - 1/2(7) + 1/2(1) + 1 = 9.0$$

$$C_{11}H_{19}F \quad\quad R + DB = 11 - 1/2(20) + 1 = 2.0$$

$$C_{11}H_{10}N_2 \quad\quad R + DB = 11 - 1/2(10) + 1/2(2) + 1 = 8.0$$

$$C_{11}H_{22}O \quad\quad R + DB = 11 - 1/2(22) + 1 = 1.0$$

$$C_{11}H_6O_2 \qquad R + DB = 11 - 1/2(6) + 1 = 9.0$$
$$C_{12}H_7F \qquad R + DB = 12 - 1/2(8) + 1 = 9.0$$
$$C_{12}H_{10}O \qquad R + DB = 12 - 1/2(10) + 1 = 8.0$$
$$C_{12}H_{26} \qquad R + DB = 12 - 1/2(26) + 1 = 0.0$$

and I can eliminate the $C_{11}H_{19}F$, $C_{11}H_{22}O$ and $C_{12}H_{26}$ possibilities from consideration. Why? Because m/z 77 and 51 indicate the presence of a benzene ring. The saturation index for a benzene ring is 4 (three double bonds forming one ring). Logically, any saturation index lower than 4 cannot provide a benzene ring.

Can I use this idea of having a benzene ring to further limit the possibilities? Well, m/z 77 will have the formula C_6H_5. Based upon this observation, I can eliminate $C_{11}H_6O_2$, $C_{12}H_7F$ and $C_{11}H_7P$ from consideration. Why? Well, subtract C_6H_5 from each of them and see what is left. In the first case, we would have a fragment with the formula C_5HO_2, with a saturation index of 5.5. This would seem to be much too unsaturated for such a small fragment. In the second case we would have a fragment with the formula C_6H_2F with a saturation index of 5.5 and, once again, it appears that such a fragment is much too unsaturated. In the third case, I would have C_5H_2P, with a saturation index of 5.5, and once again this appears to be much too unsaturated. For the time being, I will ignore these possibilities unless the remaining two formulae prove unworkable.

This leaves me with $C_{11}H_{10}N_2$ and $C_{12}H_{10}O$. Subtracting C_6H_5 from the first case leaves me with $C_5H_5N_2$ with a saturation index of 4.5, while subtracting from the second possibility leaves me with C_6H_5O. This second possibility is very interesting, since it appears to be another benzene ring with an oxygen attached.

Next I am ready to look at the logical losses and the ion fragment formula. Since those fragments at and below m/z 77 have already been identified as being characteristic of benzene (see Example 25), I will concentrate on m/z 115, m/z 141 and m/z 142.

$170 - 142 = 28$	C_2H_4, N_2, CO	Phenols, aldehydes, diaryl ethers, quinones, aliphatic nitriles, ethyl esters
$170 - 141 = 29$	CH_3N, C_2H_5, CHO	Aromatic aldehydes, phenols, aliphatic nitriles, ethyl derivatives, alkanes
$170 - 115 = 55$	C_4H_7	Butyl esters
$142 - 115 = 27$	C_2H_3, HCN	Aromatic amines, aromatic nitriles, nitrogen heterocycles
$141 - 115 = 26$	C_2H_2	Aromatic hydrocarbons

From the ion composition table I find nothing helpful in identifying the ions. Normally, I would use the isotopic data for the 115 and 142 peaks to calculate a formula for these peaks, but in this example the $A + 1$ peaks for each line amount to about 17% relative intensity, and therefore the calculations would not yield very much useful data.

The loss of 26 indicates that I have an aromatic hydrocarbon portion of the molecule, but since I have already identified a benzene ring as part of the molecule this observation provides little additional information. The loss of 27 is consistent with an aromatic nitrile or a nitrogen heterocycle. I can probably eliminate the case of a simple aromatic amine. Clearly, m/z 115 is formed from some intermediate ion and not directly from the parent. The pair of lines at m/z 141 and 142 is interesting. The loss of 28 is common for a variety of compounds, but from the ones listed we can probably eliminate everything except the diaryl ether. The m/z 141 ion is probably formed by loss of hydrogen from m/z 142.

Overall, the data are most consistent with the idea that the spectrum is that of a diaryl ether. A simple IR spectrum of the compound would probably be sufficient to rule out the possibility that the compound was a nitrogen heterocycle or a nitrile. I would identify this compound as diphenyl ether.

Compound's actual identity: diphenyl ether.

Unknown No. 19 solution

This spectrum shows significant fragmentation in the low mass end and so I suspect the compound is more aliphatic than aromatic in nature. There are two significant odd-electron peaks (other than the parent) at m/z 56 and 90. The parent ion (supposed) is at m/z 146. There does not appear to be any chlorine or bromine, and based upon the fragmentation I can also eliminate iodine.

Re-normalizing the data for m/z 146–148 and inserting the proper error limits I get the following values:

m/z	% Intensity
146	100.0
147	10.3 ± 1.0
148	4.9 ± 0.5

which indicates the presence of a sulfur atom and eliminates silicon as a possibility. Since ^{33}S has a natural abundance of 0.8, I need to subtract 0.8 from the intensity of the 147 line and recorrect.

m/z	% Intensity
146	100.0
147	9.5 ± 1.0

This indicates that there are either eight or nine carbons, one sulfur and possibly one oxygen (since there would still be intensity at the 148 line once I have accounted for the sulfur atom).

If I have eight carbons and one sulfur, I have accounted for 128 amu, leaving 18 amu for other elements. This eliminates phosphorus and fluorine (based on mass), nitrogen (based on the nitrogen rule) and effectively eliminates oxygen, since the resulting molecule would be highly unsaturated and would not be consistent with the significant fragmentations seen in the fingerprint region. The formula for this possibility would be $C_8H_{18}S$ and I would have a thioether (sulfide) or a thiol.

If I have nine carbons and one sulfur, I have a mass of 140 amu, and the formula must be C_9H_8S. The saturation index for this compound is $9 - 1/2(8) + 1 = 6.0$, and I can probably eliminate this possibility. Once again, this indicates a highly unsaturated molecule and is at odds with the nature of the spectrum.

What have I learned so far? Well, I know that I have an eight carbon sulfur compound, which is either a thioether or a thiol. The saturation index for my compound is zero

$(8 - 1/2(18) + 1 = 0.0)$ and so m/z 146 can be the parent ion. The only important question left is 'What is the structure of this compound?'.

The next thing that I have to do is to look at the logical losses.

$146 - 117 = 29$ The only loss that makes sense (given the formula) is C_2H_5. Therefore, I know that I have an ethyl group somewhere in the molecule

$146 - 103 = 43$ The only loss that makes sense is C_3H_7. Therefore, I know that I have a propyl group somewhere in the molecule

$146 - 90 = 56$ $CH_2 = CHCH_2CH_3$, $CH_3CH = CHCH_3$, or 2 CO Pentyl ketones, Ar-OBu ethers. I can eliminate the possibility of losing 2 CO. I don't have either a ketone or an ether, but thioethers are very similar to ordinary ethers in their reactions, so perhaps I do have a thioether with butyl on one side

                       Another possibility is that m/z 90 results from loss of hydrogen from m/z 91. In this case, $146 - 91 = 55$ which is consistent with C_4H_7, and once again I have a butyl group

$146 - 61 = 85$ Nothing common

$146 - 56 = 90$ Nothing common

$117 - 103 = 14$ Not a logical loss. I remember that even ordinary alkanes do not lose 14, instead they lose 28

$117 - 90 = 27$ Loss of C_2H_3 makes sense

$117 - 61 = 56$ See above

$117 - 56 = 61$ C_2H_5S. Commonly seen in thiols or thioethers

$90 - 61 = 29$ C_2H_5 is likely

$90 - 56 = 34$ H_2S is likely

$61 - 56 = 5$ Not a logical loss

Next I consult my table of ion fragment formulae. For m/z 90, I don't find anything listed. For m/z 56, I find $C_4H_8^+$. I also notice that m/z $56 + m/z$ $90 = m/z$ 146; and I am convinced that these two fragments represent the molecule basically 'cut in half'.

From the neutral loss and ion composition tables, I still have not been able to determine the type of compound. What do I do now? Well, perhaps I should review the chapter on sulfur compounds (Chapter 11). Reviewing this chapter I find the following information:

1. Thiols are characterized by $(M - 34)$ and $(M - 62)$, which is loss of H_2S and loss of H_2S followed by loss of C_2H_4. In the present example, this would result in fragments at m/z 112 and m/z 84 respectively. Neither of these fragments are present in the spectrum. It doesn't seem likely then that this is a thiol. What about a thioether? And which thioether? Well, if my observations about losing a butyl group are correct, then a dibutyl thioether would be a reasonable starting point.

2. In the spectra of thioethers (sulfides) we would expect to see α-cleavage producing $RS^+ = CH_2$, with Stevenson's rule telling us that cleavage in the longest chain is favored. In our spectrum, we have a peak at m/z 103 that is consistent with this mechanism and is also consistent with the idea of a dibutyl thioether. If the R group contains a β-hydrogen, the next step would be transfer of this hydrogen to the sulfur and ejection of an alkene, forming $CH_2 = SH^+$ at m/z 47. In the spectrum I do observe a peak of 14.5% intensity at m/z 47.

In the spectra of thioethers, we see that the RS^+ ions are more stable than the corresponding RO^+ ions, and therefore RS^+ ions will have relatively high intensities in the series *m/z* 47, 61, 75 etc. In our spectrum, we have a relatively low intensity peak at *m/z* 89 (9.4%).

Another important characteristic of thioethers is that one of the alkyl groups will transfer a β-hydrogen to the sulfur, producing a fragment 1 amu higher than the RS^+ ion (or the $R'S^+$ ion). The ion produced by β-hydrogen transfer is an OE ion, and is very useful in determining the position of the sulfur. Sure enough, the peak at *m/z* 90 is consistent with this transfer of a β-hydrogen, forming the OE ion.

So, in conclusion, I would identify this compound as a dibutyl thioether. I am not completely sure whether or not either of the butyl groups is branched.

Compound's actual identity: di-n-butyl sulfide.

Unknown No. 20 solution

Looking at this spectrum, I am immediately impressed by two very important features. The first feature is that this spectrum is of a compound with a considerable amount of aliphatic character. The second feature is that the spectrum is dominated by even mass fragments at *m/z* 42, 56, 70, 84 and 112. Could this compound be some sort of straight chain hydrocarbon with two nitrogens attached? Perhaps something like a diamine?

The high mass end of the spectrum (*m/z* 146–149) indicates that there is no chlorine or bromine, and reasonably there can't be any iodine. Re-normalizing the data and inserting the proper error limits I get the following values:

m/z	% Intensity
146	100.0
147	11.3 ± 1.1
148	4.6 ± 0.5
149	0.4 ± 0.2

If *m/z* 146 is the parent, then the $A + 2$ line indicates the presence of a sulfur atom. Correcting the intensities for the presence of a sulfur atom gives me the following values:

m/z	% Intensity $(-S)$
146	100.0
147	10.5 ± 1.0
148	0.2 ± 0.2
149	0.4 ± 0.2

I may or may not have an oxygen present in the molecule. The intensity of the $A + 3$ line (*m/z* 149) is easily explained by the presence of multiple carbons (more than one ^{13}C present in the molecule) and the presence of the sulfur isotopes.

The normalized $A + 1$ intensity indicates that there are either nine or ten carbons present, but logic tells me that this is unlikely. If I had nine carbons and one sulfur, I would have a mass of 140. This leaves 6 amu (as hydrogens) and I would have a very unsaturated molecule which would be generally inconsistent with the pattern in the low mass end of the spectrum. It seems that the largest number of carbons that I can have is eight, leaving me with only 18 amu. The most likely formula for this compound is then $C_8H_{18}S$.

So, it is apparent that I have either a thioether or a thiol. Which is it?

It seems clear that this compound is a thiol. The peaks at m/z 112 (corresponding to $M - 34$) and m/z 84 (corresponding to $M - 62$) are characteristic of thiols. The remaining even mass odd-electron ions (at m/z 42, 56 and 70) could be produced by ejection of small neutral molecules from either of these higher mass fragments.

What I have here is a linear (based on the regular pattern of the low mass end of the spectrum) thiol. I am not sure of the particular isomer, but this is definitely an octanethiol.

Compound's actual identity: 1-octanethiol.

Bibliography

The following bibliography is meant to be representative and not exhaustive.

General

Introduction to Mass Spectrometry 3rd Ed. J.T. Watson, Lippincott-Raven, Philadelphia, 1997.

Mass Spectrometry: Principles and Applications E. de Hoffmann, J.J. Charette, V. Stroobant, John Wiley & Sons, Chichester, and Masson, Paris, 1996.

Mass Spectrometry for Chemists and Biochemists R.A.W. Johnstone and M.E. Rose, Cambridge University Press, Cambridge, 1996.

Practical Organic Mass Spectrometry: A Guide for Chemical and Biochemical Analysis 2nd Ed. J.R. Chapman, John Wiley & Sons, Chichester, 1995.

Mass Spectrometry (Modern Analytical Chemistry Series) D.M. Desiderio (Editor), Plenum Press, New York, 1994.

Experimental Mass Spectrometry (Topics in Mass Spectrometry Series) D.H. Russell, Plenum Press, New York, 1994.

Mass Spectrometry (Ellis Horwood Series in Analytical Chemistry) E. Constantin, A. Schnell, R.A. Chalmers, A. Pape, Ellis Horwood Ltd., Chichester, 1991.

Mass Spectroscopy H.E. Duckworth, R.C. Barber, Cambridge University Press, Cambridge, 1990.

Mass Spectrometry (Analytical Chemistry by Open Learning Series) R. Davis, M. Frearson, F.E. Prichard (Editors), John Wiley & Sons, Chichester, 1987.

Other Types of Ionization or Detection

Practical Aspects of Ion Trap Mass Spectrometry (Modern Mass Spectrometry Series) R.E. March, J.F.J. Todd, CRC Press, Boca Raton, FL, 1995.

Fourier Transform Mass Spectrometry: Evolution, Innovation and Applications M.V. Buchanan, American Chemical Society, Washington, DC, 1987.

Lasers and Mass Spectrometry (Oxford Series on Optical Sciences) D.M. Lubman, Oxford University Press, New York, 1990.

Secondary Ion Mass Spectrometry B.L. Bentz, Elsevier, Amsterdam, 1995.

Secondary Ion Mass Spectrometry R.G. Wilson, F.A. Stevie, C.W. Magee, John Wiley & Sons, Chichester, 1989.

Secondary Ion Mass Spectrometry: Principles and Applications (International Series of Monographs on Chemistry) J.C. Vickerman, A. Brown, N.M. Reed, Oxford University Press, Oxford, 1989.

Field Desorption Mass Spectrometry (Practical Spectroscopy Series) L. Prokai, Marcel Dekker, New York, 1990.

Inductively Coupled and Microwave Induced Plasma Sources for Mass Spectrometry (RSC Analytical Spectroscopy Monographs) E.H. Evans, Royal Society of Chemistry, Cambridge, 1995.

Handbook of Inductively Coupled Plasma Mass Spectrometry K.E. Jarvis, A.L. Gray, R.S. Houk, Blackie, Glasgow and Chapman and Hall, New York, 1992.

Applications of Plasma Source Mass Spectrometry G. Holland, A.N. Eaton, Royal Society of Chemistry, Cambridge, 1991.

Chemical Ionization Mass Spectrometry 2nd Ed. A.G. Harrison, CRC Press, Boca Raton, FL, 1992.

Time-of-Flight Mass Spectrometry (ACS Symposium Series) R.J. Cotter, American Chemical Society, Washington, DC, 1994.

Time-of-Flight Mass Spectrometry and Its Applications E.W. Schlag, Elsevier, Amsterdam, 1994.

Electrospray Ionization Mass Spectrometry: Fundamentals, Instrumentation and Applications R.B. Cole, John Wiley & Sons, New York, 1997.

Electrospray: Theory, Instrumentation, Method (Journal of the American Society for Mass Spectrometry Series), Elsevier, New York, 1993.

Chromatography and Mass Spectrometry

Gas Chromatography and Mass Spectrometry: A Practical Guide F.G. Kitson, B.S. Larsen, C.N. McEwen, Academic Press, San Diego, CA, 1996.

Techniques for the Gas Chromatography–Mass Spectrometry Identification of Organic Compounds in Effluents R.E. Clement, V.Y. Taguchi, Environment Ontario, Toronto, 1991.

Introduction to Bench-Top GC/MS J.M. Halket, M.E. Rose, HD Science, Stapleford, Nottingham, 1990.

Gas Chromatography–Mass Spectrometry: A Knowledge Based User Guide F.A. Settle, M.A. Pleva, Elsevier, Amsterdam, 1988.

Basic Gas Chromatography–Mass Spectrometry: Principles and Techniques F.W. Karasek, R.E. Clement, Elsevier, Amsterdam, 1988.

Gas Chromatography/Mass Spectrometry (Modern Methods of Plant Analysis Series) H.F. Linskens, J.F. Jackson, R.S. Bandurski, Springer-Verlag, Berlin, 1986.

Liquid Chromatography/Mass Spectrometry: Techniques and Applications (Modern Analytical Chemistry Series) A.L. Yergey, Plenum Press, New York, 1990.

Interpretation of Spectra

Interpretation of Mass Spectra 4th Ed. F.W. McLafferty, F. Turecek, University Science Books, Mill Valley, CA, 1993.

Mass Spectral Interpretation C. Fenselau, J. Huley Associates, Boca Raton, FL, 1988.

Spectrometric Identification of Organic Compounds 4th Ed. R.M. Silverstein, G.C. Bassler and T.C. Morrill, John Wiley & Sons, New York, 1981.

Applications

Mass Spectrometry for Biotechnology G. Siuzdak, Academic Press, San Diego, CA, 1996.

Mass Spectrometry in the Biological Sciences A.L. Burlingame, S.A. Carr, Humana Press, Totowa, NJ, 1996.

Biochemical and Biotechnological Applications of Electrospray Ionization Mass Spectrometry (ACS Symposium Series) A.P. Snyder, American Chemical Society, Washington, DC, 1995.

Mass Spectrometry: Clinical and Biomedical Applications. Volume 2. D.M. Desiderio, Plenum Press, New York, 1994.

Mass Spectrometry of Biological Materials C.N. McEwen, B.S. Larsen, Marcel Dekker, New York, 1990.

Mass Spectrometry in Biomedical Research S.J. Gaskell, John Wiley & Sons, Chichester, 1986.

Protein and Peptide Analysis by Mass Spectrometry (Methods in Molecular Biology Series) J.R. Chapman, Humana Press, Totowa, NJ, 1996.

Mass Spectrometry of Peptides D.M. Desiderio, CRC Press, Boca Raton, FL, 1991.

Applications of Mass Spectrometry to Organic Stereochemistry J.S. Splitter, F. Turecek, VCH, New York, 1994.

Forensic Applications of Mass Spectrometry J.T. Cody, R.L. Foltz, W.A. Baumgartner, J. Park, W. Bertsch, G. Holzer, T.O. Munson, D.D. Fetterhof, J.L. Brazier, J. Yinon, CRC Press, Boca Raton, FL, 1995.

Forensic Mass Spectrometry J. Yinon, CRC Press, Boca Raton, FL, 1987.

Applications of Mass Spectrometry in Food Science J. Gilbert, Elsevier Applied Science, London, 1987.

Applications of New Mass Spectrometry Techniques in Pesticide Chemistry J.D. Rosen, John Wiley & Sons, New York, 1987.

Mass Spectrometry in Environmental Sciences F.W. Karasek, O. Hutzinger, S. Safe, Plenum Press, New York, 1985.

Topic Index

Index of Compounds and Spectra